P9-DXE-995
WITHDRAWN
JUN 05 2024
DAVID O. McKAY LIBRARY
BYU-IDAHO

TH 375 .C57 1984
Clarke, R. H.
Site supervision

SITE SUPERVISION

R.H. CLARKE

Thomas Telford Ltd · London
1984

Published by Thomas Telford Ltd, Telford House
26–34 Old Street
London EC1P 1JH

British Library Cataloguing in Publication Data

Clarke, R.H.
Site supervision
1. Building sites—management
I. Title
624'.068 TH438

ISBN 0 7277 0200 9

Set in Compugraphic Paladium in 11 on 12 pt by
MHL Typesetting Ltd, Coventry
Printed and bound in Great Britain by
Biddles Ltd, Guildford

Contents

Introduction

Civil engineering is about getting things done, and it is on construction sites that the real doing takes place.

Of course, the construction phase is the last of a series of interrelated activities which starts with the feasibility study and continues through survey, economic appraisal and site investigation to detailed design and contract preparation. But it is the site staff who must take the results of that process and make them fit into the realities of the natural and commercial environment—this is sometimes no easy task. Because theirs is the final contribution, it is the most important: there is no-one to come after and correct *their* mistakes.

That is a biased view. Biased because, like many other engineers, I have enjoyed the freedom of site life and the thrill of seeing immediate and concrete (no pun intended) results for my efforts. For engineers like me the arid phrase 'job satisfaction' is quite inadequate to convey the excitement and fulfilment which comes from the successful completion of a long concrete pour, the sight of heavy plant working under floodlights or the sound of scrapers taking their first bite from a big cutting at the start of the earthworks season.

Even those of a less romantic disposition, and not all engineers have either the temperament or inclination for a career on site, must recognise the importance of site work, not only as the end product of most engineering activity but also as an essential part of professional training. The traditional, and still the most heavily trodden, path to the practical experience required for qualification as a civil engineer lies through construction sites. Thus, although later in their careers many engineers choose to specialise in design, research and development, geotechnics or one of the other branches of the profession, almost all will spend some significant part of their working lives with mud on their boots. And for that large majority, over 60 per cent, who work for government (local and national), public authorities and consultants, that time will be spent in the supervision of works carried out by contractors.

It is a world very different from the drawing office or the laboratory, one for which most newcomers will have had little formal preparation. Site supervision is, after all, more of an art (or at least a craft) than a science, and does not lend itself well to classroom tuition or textbook treatment. It is, however, too expensive an activity for site staff to learn by their own mistakes and too fast-moving for them to be taught by 'sitting next to Nellie'. So this book, although not a textbook, sets out some guidance on how the basic principles of contract administration can be translated into the practicalities of effective site supervision. I say 'can' and not 'should' or 'must' because these are the views of one resident engineer on what constitutes good practice. My objective is to provide sufficient explanation, description and comment to allow any newcomer, and a few old hands, to think their own way through the problems of supervising a contract.

To assist in this aim I have included as many examples as possible. Some are taken from the reported facts of court cases and so indicate how the law views certain aspects of contracting and construction. Each of these is identified by the title of the case, the year in which it was decided and the reference of the corresponding Law Report. As with every legal judgement these decisions are based on the circumstances of individual cases, but they are quoted because of the general principles which they also establish.

The majority of the examples are not formally identified, although they are all real. I have not given the names of the parties involved nor any other details which would make the contracts clearly recognisable for two reasons:

(1) Many examples relate to disputes which were settled on site or by arbitration and were thus never in the public do-

main and I am sure that most of the employers, contractors and Engineers involved would not want me to record their private disagreements in print.

(2) In some examples, irrelevant details have been omitted or simplified in the interests of clarity and brevity and it therefore would be misleading to be specific about time and place in respect of such 'edited' descriptions.

The examples are intended to illustrate concepts, to act as a stimulus to thought and not as a substitute for it. No two situations are exactly alike and each must be treated on its merits, although within certain guiding principles. It is not my intention to provide a source of 'instant answers', always available to be thrust under the agent's nose at the first sign of disagreement. I would be sorry indeed to see this book, or any other, used in that way, for the only result would be to drive a wedge between the agent's staff and the resident engineer's team. The gap of understanding between them needs to be narrowed and not widened, and this can only be achieved if site staff understand what supervision is about and what employers and contractors expect from the contract.

Anyone coming new to site work will be daunted by the responsibilities and tasks which the resident engineer has to shoulder. To get a proper perspective it is worth remembering two things. Although the resident engineer is in charge of supervision, he does not work alone but with the support of a team whose energies and skills will, if properly directed, allow him to be everywhere at once, to always keep up to date and to have all the skills necessary to administer the contract. Secondly, no-one can expect the resident engineer to respond to every problem with an immediate solution, and on some occasions the right answer will be, 'I don't know, but I'll find out'. It is trite but nevertheless true that engineers, especially those who work on site, never finish learning. That learning process (and perhaps the self-confidence of our profession) would benefit if engineers were more willing to share their experiences and opinions. I have tried sharing some of my own in the hope that they will be of use to practising, and practical, engineers. After all, this book was born and bred on site and it is on site that I hope it will be used.

Chapter 1
The contract

Civil engineering works are carried out by construction firms employed under a form of agreement known as a contract. The obvious risks of over-simplification can be balanced by the benefits of clarity and so it will be useful to bear in mind through everything that follows that the contract is about three things: the employer, the contractor, and their money.

Tendering for a contract

Most civil engineering work is carried out under contracts which are awarded as a result of competitive bidding between a group of firms selected on the basis of their suitability for the particular project. The process of tendering and the techniques of tender evaluation are too complex for exhaustive treatment here but an understanding of the basic principles is an essential background to the business of site supervision.

The usual starting point is the issue by the person or organisation promoting the project of a set of tender documents to each member of the chosen group of bidders. The documents comprise the drawings, specification and schedules which describe the project in detail and also a statement of the general provisions upon which any subsequent contract would be based. In addition, bidders are normally supplied with 'Instructions for Tendering' which provide details of the date, place and method for the return of the bids, which usually must be delivered in unmarked sealed envelopes to prevent identification of, and interference with, any tender prior to opening. Typical 'Instructions' would also include information on the likely delay between return of the tenders and award of the contract, procedures for dealing with queries or requests for further data during the tender period and a proviso to the effect that the promoter is not bound to accept 'the lowest or any other tender', thus allowing the project to be shelved without the risk of paying damages to a tenderer. Often bidders are also advised that no 'qualified tenders' (i.e. those with conditions attached) will be accepted and that 'alternative tenders' (i.e. offers to construct the works which incorporate some significant change from the original design) will only be considered if accompanied by a tender based on the documents as issued.

A very low tender which carried a covering letter explaining that the price for disposing of surplus material from mass foundations was conditional upon the success of a planning application to fill an old chalk pit would be regarded as 'qualified' and, if the instructions banned such tenders, unacceptable. By contrast, if a completely unconditional tender for the same job were accompanied by a second, and cheaper, alternative based upon the use of specialist ground treatment to reduce the size of the foundations and so eliminate the need to dispose of surplus excavated material, both would be acceptable and open to consideration by the promoter.

It must always be remembered that the instructions for tendering are not part of the formal contract which will subsequently come into existence, and anything considered vital to that agreement must therefore be put into the body of the tender documents to have proper effect.

The tender period is very short by comparison with the design or construction stages, eight to twelve weeks is typical even for major projects, and the bidders will be working against a very tight schedule to complete their estimates in time. At this stage the quantifying of risks and uncertainties becomes an inevitable and essential exercise, and one which is likely to be seen from different viewpoints by employer and contractor at a later date. The employer will rely on the 'sufficiency' of the tender on which the contract will be based; put simply this means that the contractor is considered to have examined all aspects of the project, including the site and its surroundings, and taken all the usual risks into account in drawing up his bid which is then considered as fully inclusive. On the other hand the contractor will argue that the limited time available for investigation and the competitive nature of the bidding force him into accepting the information on the contract documents as being conclusive, and that anything not immediately apparent from an examination of that information is 'unforeseen' and must incur an extra cost.

More will be heard of these two arguments later, but they can fairly be said to provide the basis for a large proportion of all contractual disputes, particularly in respect of ground conditions. Contractors have a valid point. The winning tender is generally the lowest one and it might therefore appear foolish to go behind or beyond the information provided by the promoter to look for trouble which is only likely to increase the bid. The promoter has decided upon competitive tendering and must take the risks which go with its advantages. This argument, although attractive, must be considered in the light of the proposed contract and the state of mind of the two parties:

(1) The promoter is buying experience and expertise, and construction companies purport to be capable of undertaking work in a business which is known to involve a high degree of uncertainty.
(2) The data on the state and condition of the site which promoters provide is rarely unqualified or guaranteed to be comprehensive and in most civil engineering contracts the contractor is taken to have inspected the site before signing.
(3) There is a procedure for putting queries or requests for more information during the tender period and alternative tenders are acceptable if properly submitted.

The rules may be hard, but they *are* the rules.

It would be unwise to suppose that this summary is the beginning and end of the debate, but it does emphasise the care that must go into preparing tender documents and answering queries during the tender period in order to minimise the very real difficulties faced by the bidders. In particular it must be remembered that information given to a tenderer, especially in respect of ground conditions, will become binding on the employer in any subsequent contract. It is therefore only fair that any queries which elicit information relevant to the bid but which are not included, or not included clearly, in the tender documents should be answered in writing and sent to all the tenderers.

When the due date arrives, the bids are opened, checked for arithmetic accuracy and subjected to a process of evaluation to ensure that they are 'clean tenders' and not qualified or based upon obviously unreasonable or inconsistent rates. The tender documents, as amended by any formal supplements issued during the tender period, can then become the 'Contract Document', the promoter becomes the 'Employer' and the successful tenderer the 'Contractor'.

Types of contract

Contracts are traditionally divided into three categories: lump sum, cost-plus and measure-and-value. They differ in the way in which the contractor is paid, but all three are executed in the orthodox manner with the employer arranging for the design and supervision of the works and the construction being undertaken by the contractor. Occasionally projects are tackled through 'package deals' in which the contractor is responsible for both design and construction phases, but this is relatively rare. A new and somewhat ill-defined variation on this theme is the 'management contract' in which a project, the design of which may or may not have involved the contractor, is administered by a specialist firm who in turn award various subsidiary contracts and supervise their execution for a fee based on a number of complex value-related factors. This use of a contractor to supervise contractors contains an element of 'poacher turned gamekeeper' and depends very heavily on the ability of the poacher not only to remember the tricks of his own trade but also to assume the values of the gamekeeper's calling: although it has gained some popularity in the building field, this type of contract is almost unknown in civils work.

The distinguishing features of the main categories are described below.

Lump sum
This is the pure fixed-price contract in which the agreement is to complete the entire works for a predetermined amount which represents the limit of the contractor's entitlement. Anything considered 'indispensably necessary' to the required end product is considered to be included, even if omitted from the contract documents, and in extreme cases the lump sum may be assumed to cover all risks including errors in the details presented in those documents.

Thus, in the notorious San Paulo Railway case of 1873 the agreement was to lay a railway between two specified points for a fixed price. Errors in the engineering drawings resulted in almost double the stated volume of excavation being required but this did not entitle the contractor to any extra payment. Needless to say, lump sums are viewed with disfavour by contractors, particularly if combined with competitive tendering, and are not widely used in modern civil engineering work.

Cost-plus
In this form of contract, the price is determined on the basis of the cost of labour, plant and materials plus a percentage fee (or a fixed sum) to cover overheads and profit — tenderers compete on this 'plus' element only. The arrangement is obviously favoured by contractors, but is less attractive to those paying for the work, who are likely to see it as an open-ended commitment. To provide incentives for economy, a number of variations have been developed such as the 'target cost' contract in which a preliminary cost is estimated and on completion the difference between this target and the actual cost forms the basis for calculating an adjustment, up or down, to the contractor's fee according to some pre-arranged formula. Cost-plus is often used as the basis for especially difficult or risky projects or for work of an unusual nature where accurate estimating would be impossible.

Measure-and-value

The wording of the Fifth Edition of the Institution of Civil Engineers' Conditions of Contract specifically places this widely-used standard form of contract in the measure-and-value category. In this type of agreement the figures inserted by the tenderer against the various items in the bill of quantities form a schedule of rates, the contract sum is thus calculated on completion by measuring the work done and valuing it according to those rates. The tender total, arrived at by multiplying the rates by the estimated quantities, is thus no more than a device for identifying the lowest overall bid and has no large contractual significance.

Since the escalation of oil prices in 1973 and the presence of relatively high levels of inflation in our economic system, contracts have invariably included some form of adjustment for the variation in costs of basic materials and labour during the life of the project. These variation-of-price contracts rely on the use of specified official index figures (commonly called the 'Baxter Index') to determine a 'price fluctuation factor' which relates to a 'base date' six weeks before the contract was awarded. The appropriate factor is calculated at each monthly valuation of the work done and is used to adjust the sum due to the contractor as determined from his tender rates. A modern variation-of-price, measure-and-value contract is thus at the opposite end of the spectrum to that represented by the fixed-price, lump sum agreement which governed the construction of the San Paulo Railway over a century ago and caused its contractor such grief.

The basis of a contract

A contract is an agreement enforceable at law. All sorts of transactions, major and minor, fall into this category, and the purchase of a railway ticket is just as much a contract as is an agreement to construct a motorway or a power station.

Construction contracts are thus part of a very wide ranging branch of the law.

When can an agreement be called a contract? English law recognises four basic requirements:

(1) offer (i.e. the contractor's tender for the work);
(2) acceptance (i.e. the employer's award to the successful tenderer);
(3) consideration (the contribution of each party to the bargain, i.e. the contractor's undertaking to do the work in return for the employer's promise of payment);
(4) intention to be legally bound.

The use of standard procedures and documents for the tendering process and the involvement of legal advisers makes it highly unlikely that any of these requirements would fail and thus allow a project to start without the existence of a valid contract. Even if such an eventuality were to occur, it can safely be assumed that employer and contractor — and their lawyers — would move swiftly to bring a proper contract into being, complete with all the necessary safeguards to cover the initial error. The staff on site to supervise the project would probably not learn about it, if at all, until long after the event.

It may appear that site staff do not need to know anything about the basic principles of the law of contract, and can leave such matters to administrators and lawyers, but this would be a dangerous division of labour. Although it is unlikely that they will ever be troubled by the form of the contract, they will certainly need to understand all about its content: the extent of the work which the contract covers. The contract documents may be very comprehensive, but the actual application of what is stated, and in some cases implied, remains subject to interpretation, often at site level.

You will search in vain for a 'Contracts Act' or any legislation which sets out the law on the administration of contracts. There is no equivalent of the Highways Act or the Health and Safety at Work Act because the law of contract, unlike, for instance, administrative law and employment law, does not rest on statutes but is primarily built on case law. Almost every important principle of contract law is derived from the recorded decision of a judge who has pronounced his verdict on the facts of a particular case. It is therefore in the full records of the Law Reports and the abridged notes in commercially published casebooks and textbooks that the basic information will be found, but in each instance the decision will be in respect of a unique set of events and its application to any new situation will be a question of judgement. To an engineer, this approach will seem alarmingly imprecise and the Common Law of England may appear something less than the 'main pillars and supporters of the fabric of the commonwealth'. The system has its advantages, however, being flexible and responsive to change, and the complexities which are bound to stem from a large and growing body of case law are considered by most legal commentators to be a reasonable price to pay for these benefits. For full appreciation of the theory and practice of contract law it is necessary to refer to one of the expert, and deservedly famous, publications which cover the field in depth and with a degree of detail which are impossible here. For those whose interest is kindled, the law can be a fascinating and stimulating study: the two examples of case law which follow are presented firstly to explain, by illustration, the extent of real and implied contract terms and secondly, to give the flavour of a legal reference text.

In every contract, no matter how detailed the documentation, certain requirements will be left unmentioned, often because they were considered to be 'self-evident'. These are the *implied* terms and the principle of how such terms can be recognised was established in a shipping case commonly known as 'The Moorcock', which can be looked up under the Law Report reference (1889) 37 W.R. 439. The Moorcock was a barge which tied up at a jetty on the tidal reaches of the Thames. The ship's owners had a contract with the owners of the jetty which allowed them to discharge her cargo on payment of certain landing fees, both parties knowing that the vessel would ground at low tide. There was a hard ridge beneath the mud and the Moorcock suffered damage. Her owners sued for the cost of the repairs and despite there being no guarantee of the safety of the mooring in the contract they were successful. The judges (the case went to the Court of Appeal) implied an undertaking by the jetty owners that the river bottom was, so far as reasonable care could provide, in such a condition as not to endanger the grounded ship. In a later case involving an implied term, Lord Justice Scrutton defined it as a term which '. . . if at the time the contract was being negotiated someone had said to the parties, "What will happen in such a case?" they would both have replied: "Of course so-and-so will happen; we did not trouble to say that; it is too clear"'.

The plans for a steel lifting bridge were drawn up under a lump sum 'design and build' contract, with the stipulation that the bridge deck could be supported by only one of its two hydraulic rams in the event of failure. During construction it was found that expensive modifications had to be made because the deck distorted to a damaging degree when unevenly supported on a single ram. There is an implied term in this contract which requires the designer to provide not only a ram system powerful enough to hold up the deck in the event of a partial failure but also a deck sufficiently rigid to remain sound in this condition: without the second, implied, term, the first requirement is largely useless. There can, therefore, be no extra payment.

It must not be imagined that every missing provision can be read into a contract as an implied term, nor that a party to a contract can be held bound by conditions, however obvious they may appear to be to the other party, if they lie outside the contract. Implied terms must be genuinely self-evident. Now that specifications, methods of measurement and standard details have become so comprehensively drafted it is very risky to rely on implication as an alternative to presenting clear requirements and full information in the contract documents. This hard lesson is best illustrated by another shipping case, McCutcheon *v* David MacBrayne Ltd. – the Law Report reference is (1964) W.L.R. 125 which concerned the validity of a set of trading conditions. (Much case law in contract comes from the activities of railway and shipping companies, who are, or were, both apparently keener to take disputes to court than civil engineering firms).

Mr. McCutcheon lived on the Isle of Islay and wanted to send his car to the mainland by one of David MacBrayne's steamers. He asked his brother-in-law, who rejoiced in the name of McSporran, to make the arrangements and he duly went to the shipping line's local office, paid the charge and took a receipt which stated that 'all goods were carried subject to the conditions set out in the notices'. These 'notices' were a set of 27 conditions displayed on the walls of the office and reprinted, in very small type, on a document called a 'risk note' which customers were asked to sign. Although Mr. McSporran had sent cargo many times and had always signed the 'risk note', on this occasion he was not asked to sign. All the previous cargoes had arrived safely, but this one did not — the ship struck a rock and the car was lost. MacBraynes relied on one of their conditions which excluded liability for any loss 'wheresoever or whensoever occurring' and said that Mr. McSporran had signed so many risk notes in the past that he must be deemed to have known the trading conditions. The House of Lords ruled otherwise: whether or not Mr. McSporran had previously read and known the conditions (in fact he professed ignorance of them) was irrelevant; all that mattered was that in this particular agreement the conditions had not been put forward for the customer's signature thus making it a straightforward contract to carry the car to the mainland without any implied exclusion clauses. The shipping company had to pay damages for the loss.

On a large industrial development project a 600 mm water main ran through the site and had to be diverted. The tender documents showed the line and level of the main, its size, the route of the diverted section and pointed out the need to provide facilities for the Water Board's own men to come in and carry out the work. The designer of the project must have felt that the impossibility of shutting off such a large main for more than a very short time, and the consequent need for the diverted section to be completed before the old length could be removed, was obvious, so obvious that nothing was put in the documents. Should it have been obvious to the tenderers too? The contractor subsequently asserted that his programming and estimating had been based on the assumption that the supply could be shut off, allowing the original main to be abandoned before the new section was built, any other sequence being more disruptive and thus more costly. Even if it had been the general practice of the Water Board to permit only the briefest interruption to large mains, the failure to include specific reference to such an important restriction makes the employer liable for the reasonable extra costs incurred by the contractor in re-phasing his work.

It should now be clear that the contents of the contract can be a very real concern for the site staff, although the dividing line between what is in and what is out may still seem difficult to discern. The important thing to remember is that a term can be implied if it is necessary to give the contract what has been called 'business efficacy'; in other words, to make it a realistic and workable agreement based on normal practice. No-one would consider it a 'normal' contract to design a lifting bridge which could not survive the action of the fail-safe arrangement in the hydraulic system, but a contract which does not specify the phasing of a particular series of operations is still a workable agreement even if one party is not satisfied with the consequences of the omission.

What has been said above refers to the position under English Law. For legal purposes, Scotland is a foreign country with its own courts, judiciary and procedures, although subject to the legislative authority of the British Parliament at Westminster. Any contract intended for use in Scotland should contain a clause stating that it is to be construed and operated as a Scottish contract. This would ensure its validity, but would not eliminate the differences between the two legal systems (for instance, consideration does not have the same importance in Scottish as in English contracts and the law of negligence varies on significant points) which remain and require specialist explanation.

The ICE Conditions of Contract

A civil engineering contract can take any form, providing it satisfies the four basic requirements to make it legally enforceable. There is no legal objection to every engineering project having its own unique agreement containing specially framed terms and conditions. However, uniformity does bring significant benefits, albeit at the expense of flexibility. Estimating, for instance, becomes more accurate when direct comparisons can be made between contracts with identical provisions and tender preparation can proceed more smoothly when there are no unfamiliar terms to be examined and assessed. The attitude of contractors to the execution of the work and of resident engineers to its supervision should show more consistency than under a system of 'one-off' contracts. Perhaps the most important benefit is that the existence of a single document allows an accepted body of knowledge to be built up about its meaning and application, thus allowing gradual but continuous development so that errors and loopholes are eliminated and not repeated. These advantages have encouraged the widespread adoption of 'standard form' contracts, three examples of which account for the majority of construction work in the United Kingdom. They are:

(1) for building projects the Joint Contracts Tribunal (JCT) Conditions;
(2) for civil engineering the Institution of Civil Engineers (ICE) Conditions of Contract;
(3) for Government contracts the 'GC/Works/1'.

The Government form, being prepared unilaterally by the employing authority, is generally considered to be the clearest and best drafted and is well known for the way in which it places almost all the risks on the contractor. Both the JCT and ICE forms are the result of negotiation between representatives from both sides of the industry and reflect, in their ambiguities and a certain lack of crucial definition, the inevitable compromises and accommodations which such a process entails. These two forms have frequently been criticised as excessively legalistic and obscure. Lawyers rightly respond that a very simple and clear document could have been drawn up if the parties had really wanted one (or been able to agree upon what it should say!).

Throughout this book, unless otherwise stated, any reference to the Conditions of Contract means the ICE form. A number of detailed commentaries on the ICE Conditions of Contract are available and no attempt will be made to paraphrase those works of reference here. However, it is worthwhile setting out the clauses in the ICE Conditions of Contract which are particularly relevant to site supervision. A summary of each relevant clause follows, and chapter references are given in brackets to indicate where more detail can be found.

Clause 2	sets out the functions and powers of the engineer's representative and his assistants; provides for delegation by the Engineer; allows site decisions to be referred back to the Engineer (Chapters 2 and 3);
Clause 4	establishes the principles for sub-contracting (Chapter 8);
Clause 5	requires the contract documents to be mutually explanatory and, where there is doubt, gives the Engineer the duty to resolve any discrepancies (Chapter 4);
Clause 8	sets out the contractor's general obligation to provide everything necessary to construct the works and his particular responsibility for the safety of site operations (Chapters 4 and 7);
Clauses 11 and 12	cover the presentation by the contractor of a fully-inclusive offer as the basis of the contract, subject to a right to claim against 'unforeseen' conditions (Chapter 13);
Clause 13	appears to give sweeping powers of acceptance and approval to the Engineer, which are in fact heavily qualified (Chapter 9);
Clause 14	requires a programme to be submitted for approval and provides for its revision; empowers the Engineer to obtain and assess details of the contractor's proposed methods of working (Chapter 5);
Clause 15	lays down standards for the contractor's superintendence of the works (Chapter 9);
Clause 17	makes setting-out entirely the responsibility of the contractor (Chapter 9);
Clause 19	sets out obligations for the safety and

	security of the site, which rest chiefly, but not exclusively, with the contractor (Chapter 7);
Clause 20	requires the contractor to protect the works and make good any loss or damage unless it is caused by one of the specified 'excepted risks' (Chapters 7 and 9);
Clauses 29 and 30	restrict disturbance, nuisance, damage to public highways etc. (Chapters 5 and 7);
Clause 36	provides for testing and quality control of materials and workmanship (Chapter 9);
Clauses 38 and 39	lay down the principles for inspection of work in progress and for its removal if rejected (Chapter 9);
Clause 40	provides the mechanism by which the Engineer can suspend the progress of all or part of the works (Chapter 6);
Clause 42	sets out the contractor's right to have possession of the site (Chapter 12);
Clauses 43 and 44	provide for a fixed time for completion, subject to extension by the Engineer in accordance with a specified procedure for assessment (Chapter 6);
Clause 48	sets out the procedure for the issue of a completion certificate when the works or any specified parts of it are substantially completed (Chapter 6);
Clause 49	defines the maintenance period and lays down the contractor's obligations to carry out all repairs excepting fair wear and tear (Chapter 6);
Clause 51	empowers the Engineer, through the issue of variation orders, to make alterations, deletions or additions to any aspect of the works (Chapter 10);
Clause 52	sets out the procedure by which varia-

tion orders are evaluated by the Engineer; gives the Engineer discretion to pay daywork rates for variation orders subject to the proper presentation of accounts etc. as specified; sets out the procedure for the notification of claims and places the obligation on the contractor to keep and maintain full records (Chapters 10, 12 and 13);

Clause 54 — allows the Engineer to approve payment for goods and materials which are not on the site and sets out comprehensive safeguards to ensure that they are properly identified and cannot be claimed by any third party (Chapter 11);

Clauses 55 to 57 — lay down the principles on which the measurement and valuation of the works will be based, including provisions for correcting errors in the bill of quantities and for mutual establishment of measurement with the contractor (Chapter 11);

Clauses 58 and 59 — define provisional sum, prime cost item and nominated sub-contractor; set out the procedure by which the sub-contractor can be nominated and allow the contractor the right to object; make special provisions for payments to nominated sub-contractors (Chapters 4, 10 and 11);

Clause 60 — sets out the process by which monthly statements of accounts are submitted by the contractor for evaluation and certification by the Engineer; provides a timetable for the drawing up of the final account on completion of the works; allows for the retention of a proportion of the value of the works for subse-

	quent release on completion and after the period of maintenance has expired satisfactorily (Chapter 11);
Clause 62	empowers the Engineer to carry out urgent repairs in an emergency when the contractor is unable or unwilling to act (Chapter 7);
Clause 66	sets out the procedure by which disputes are referred to the Engineer for his formal decision and, in the event of continued disagreement, placed before an arbitrator (Chapter 13).

Regular readers of the ICE Conditions of Contract will have noticed several omissions: the clauses on insurance, for example, and on forfeiture and frustration. These are not matters which engineers can ignore, far from it, but they do represent areas into which the site staff should not venture without specialist advice. Also missing is the famous clause 65, the 'War Clause', which still fills two pages of the Fifth Edition with detailed provisions. In a standard form designed solely for use in the United Kingdom in the age of nuclear weapons and chemical warfare this almost amounts to black comedy. Should it ever be needed, there will be precious few engineers who will bother, let alone care, about its implementation.

Chapter 2
The participants

Parties to the contract

The 'Contract' is an agreement between two parties: the 'Employer', who wants the work done, promises to pay for its execution and will own it when it is finished; and the 'Contractor', who agrees to carry out the work, provides all the necessary materials and resources and will be paid for satisfactory completion.

The employer in a modern civils contract of any size is most unlikely to be an individual, but rather a large company or public authority. Nevertheless, the law invests such organisations with a 'corporate personality' and recognises that they act, through their representatives or officers, in much the same way as a person. Similarly, for legal purposes, the contractor is also treated as a person despite being in most cases a limited company. Referring to the employer or contractor as 'he', therefore, is not just a figure of speech but a fair reflection of their status under the law.

This concept of an artificial person has been developed so that 'corporations' (the correct term for organisations recognised by the law as independent personalities) might have access to the remedies of our legal system as well as being subject to its procedures and penalties. Thus a corporation can sue for damages, or defend itself against a claim of negligence; it can enter into a contract, or be summonsed; it can bring an action to recover a debt or be found guilty of a crime. The concept has certain limits, however, and not just the obvious ones that prevent a corporation marrying or committing a personal assault: a corporation can be fined for a criminal offence, but it cannot be imprisoned. This last restriction may serve as a useful reminder that individuals cannot always shelter behind the veil of 'corporate personality'. An action can be brought against contractor's and employer's personnel, and indeed against the resident engineer's staff, not only as representatives of 'corporations' but also as individuals. This is frequently done where the company or organisation has established correct procedures and the individual has abused or ignored them. Health and Safety at Work offences form a typical and extensive category.

The term 'Employer' is to some extent misleading as the association with the 'Contractor' is not in any way similar to the employer/employee relationship which is found in everyday industrial and commercial labour arrangements. In these situations the employer is part of what the law still calls a 'master and servant' relationship, which is characterised by the 'master's' right to control the activities of his 'servants' who are employed to perform services which are an integral part of the business. In the eyes of the law the contractor does not fall into this category, but has the status of an 'independent contractor'. Members of this group usually work under an agreement to perform a specific act or acts but are not, in Lord Denning's words, 'part and parcel of the organisation'. The importance of this distinction lies in the liability of the employer where harm is caused by a person doing work for him. In the 'master and servant' situation, the employer may be liable but where an independent contractor is involved, he is not.

If one of the employer's own vehicles delivering materials to the site crashes due to bad driving and injures a pedestrian then the employer may be liable for damages. If, on the other hand, the vehicle had been owned by the contractor, any liability would be entirely his, notwithstanding the fact that the vehicle and driver were doing work required and paid for by the employer.

'Independent' the contractor may be in legal terms, but he is nevertheless bound by the obligations laid down in the agreement with the employer. The ICE Conditions of Contract describe them in general terms. The detailed requirements for the contractor's work are found in the remainder of the contract documents – the drawings, specification, bills of quantities etc. – and his compliance with these, and the exercise of his general obligations, is made subject to a test of 'satisfaction'. It is not the employer who must be satisfied, however, for the two parties to the contract have agreed to pass this authority over to a third person: the Engineer.

The role of the Engineer

The Engineer is appointed and paid by the employer. Usually he is the designer and, although there is no contractual requirement for this to be so, a chartered engineer. Sometimes the Engineer is a named individual, although on contracts of any significant size it is more likely that the appointment will be impersonal with the title of some official post or the name of a partnership quoted in the contract documents: this avoids any uncertainty arising out of the the resignation or departure of a named Engineer. In any event, the wording of clause 1 of the ICE Conditions of Contract, which covers the appointment of the Engineer, allows the employer to replace him without affecting the agreement with the contractor, provided the change is notified.

The office of Engineer is usually filled in one of three ways:

(1) Employers who have their own professional experts may appoint the Engineer from amongst their staff (as when a British Rail Region appoints its Chief Civil Engineer, or a County Council its County Surveyor).
(2) Employers with a long-term or continuous programme of civil engineering projects may delegate their promotion and management to an agent with technical staff who will in turn appoint the Engineer from amongst his own employees (as when Department of Transport work is carried out by County Councils with the County Surveyor as Engineer, or when Water Authority schemes are executed by a District Council with its Chief Technical Officer as Engineer).
(3) Employers, whether in one of the above categories or not,

may choose to engage professional assistance on a short-term contract for the supervision of the construction of a particular project or related series of projects (as when a County Council, the Department of Transport, the Central Electricity Generating Board, a private developer etc. employ a firm of consulting engineers).

There is one common denominator — the Engineer must be impartial. He participates in the contract but is not a party to it for he gains no benefit from its execution (his payment is the subject of a separate agreement outside the contract). He is not the employer's representative and although in many of his duties his legal position is that of an agent empowered to act on the employer's behalf (e.g. in granting permission for the sub-letting of parts of the contract), his standards of behaviour must differ significantly from the layman's conception of how an agent can operate, for he cannot actively advance the employer's interests at the expense of the contractor. The Engineer is bound to protect the employer's rights, but this is not the same as promoting his interests and the contractor's rights must be kept clearly in view — to fail to recognise this would be quite contrary to the spirit of the contract and the letter of the Engineer's appointment.

Consultants are normally appointed as Engineer under the 'Model Service Agreement' ('Form A') drawn up by the Association of Consulting Engineers. This document sets out the Engineer's duties at the construction stage (clause 2) and further provisions cover the appointment of his site staff (clause 9) and the exercise of care and diligence in the performance of his professional duties (clause 12). This last requirement is of the utmost importance: not only is the Engineer charged to exercise 'all reasonable skill, care and diligence', but he must also 'act fairly as between the Client (i.e. the Employer) and the Contractor'. Engineers appointed directly from the staff of the employer or the employers' agent authority are rarely asked to sign any similar undertaking, nevertheless, there is no doubt that the same standards would be applied by implication to them, as well as the general common law requirement for professional advisers to display reasonable care and skill in the practice of their calling.

The Engineer's task is a difficult one as the distinction between proper involvement in the administration of the contract and biased support for the position of his employer is not always an easy one to draw in practice. The employer, unlike the Engineer and the contractor, is rarely represented on site and so it is generally true that unless the employer's side of any dispute is put by the Engineer it may not be put at all. To let the employer's case go by default in this way would be just as much a failure of the Engineer's impartiality as refusing to consider the contractor's reasoned arguments. The Engineer is not in breach of his duty to act fairly and without bias when he argues against the contractor, provided he does so with professional detachment and an open mind.

On the face of it, an Engineer whose appointment comes through the engagement of a firm of consultants should find it easier to distance himself from the employer. The pressure, conscious or unconscious, inherent in the appointment of an Engineer who is in the direct service of the employer or the employer's agent is absent and the consultant can claim a clear independence which the other forms of appointment do not exhibit. However, circumstances are not really very different: the client/consultant relationship contains as much potential for pressure as that of principal/agent and employer/employee, for a consultant will wish to keep in good standing with his client and the prospects of future work can affect judgement just as much as the prospects of promotion. There is no short cut to impartiality, the key is constant self-questioning and the willingness to submit every decision to critical examination. Every Engineer ought to bear in mind that any one of his actions or statements could, in the last resort, be reviewed before an arbitrator, or even a judge, and should conduct himself accordingly.

The concept of the 'impartial Engineer' has recently become the subject of active debate within the profession, with much concern expressed over the erosion of his authority and capacity to act fairly resulting from the involvement of quantity surveyors, auditors and accountants. It is interesting to note that the recognition of an absolute duty of 'fairness' is a relatively modern one, much reinforced by legal decisions over the last forty or fifty years, and that in the rosy past

Engineers might be better described as 'arbitrary' rather than 'impartial' in many of their actions. An extreme, but not unique, example appears in the career of Brunel. As Engineer to the Oxford, Worcestershire and Wolverhampton Railway, in the summer of 1851, he found himself involved in a dispute over the terms of the contract, the level of payments and the ownership of plant on the Mickleton Tunnel section and decided that the existing contractor should be replaced. Brunel did not trouble the employer with this problem (at least not officially) but simply marched at the head of 2,000 labourers and forcibly ejected the understandably reluctant contractor and installed a new one, ignoring at least three readings of the Riot Act by local magistrates. Only when the works had been occupied and the original contractor's labour force dispersed did the Engineer suggest arbitration. Brunel was lucky, and at the 'Battle of Mickleton Tunnel' casualties were slight—unless, of course, one includes the contract.

The duties and powers of the Engineer

The Engineer's duties and powers can be found displayed with varying degrees of clarity on almost every page of the ICE Conditions of Contract, and any complete listing would be bound to involve the reproduction, or at least the paraphrasing, of a large part of that document. What follows is no more than a review, and for convenience the chief duties and powers are assembled in three categories which themselves represent the main functions of the Engineer.

Approval and acceptance:

(1) permission to sub-let parts of the works;
(2) approval of the contractor's programme and his proposed methods of working;
(3) approval of the contractor's agent and (via the sanction of removal) of his employees;
(4) acceptance of workmanship and approval of the quality of materials;
(5) approval of measures to deal with problems such as

unforeseen physical conditions or slow progress;
(6) the issue of certificates accepting the works as substantially complete and accepting the satisfactory completion of the maintenance period;
(7) the application of the test of 'satisfaction' as the standard of compliance for all matters pertaining to the contract.

Evaluation and adjudication

(1) explanation and adjustment of ambiguities, discrepancies, errors and omissions in the contract documents;
(2) assessment of whether physical conditions or artificial obstructions were 'unforeseen';
(3) assessment of delays and evaluation of extra costs incurred by the contractor;
(4) allocation of liability for damage to the works or public highways;
(5) assessment of the rate of progress against completion date;
(6) assessment of any extension to the contract period;
(7) measurement and valuation of the works for interim payment and final account;
(8) evaluation of whether and how contract rates should be varied to take into account changes in the works;
(9) adjudication in disputes between employer and contractor.

Instruction:

(1) the issue of further drawings and variation orders to supplement, delete or modify any part of the works;
(2) the suspension of the works due to weather, default of the contractor, reasons of safety or the presence of unforeseen physical conditions or artificial obstructions;
(3) the direction of the contractor in dealing with unforeseen physical conditions;
(4) the removal of improper work or materials;
(5) the direction of the contractor regarding the use of provisional or prime cost items in the bill of quantities and the employment of nominated sub-contractors;
(6) the direction of the contractor regarding the keeping and

maintenance of particular contemporary records in connection with any claim.

In exercising this authority it is obvious the Engineer must use his personal judgement. However, he must always operate within the terms of the contract and cannot substitute his opinion of what may be fair and reasonable for a specific provision: it is the agreement between employer and contractor which comes first. In the same way as legal commentators distinguish between the dispensation of justice and the administration of the law, so can a distinction be drawn between what the Engineer may think is right in engineering terms and what the contract requires.

In former times the Engineer was seen in a 'quasi-judicial' or 'quasi-arbitral' role, independently determining the relevant facts and applying the contract to them. Engineer's decisions, even if mistaken, were made binding on both parties to the contract and where arbitration clauses were written into the agreement it was not unusual for the Engineer to be named as arbitrator. This is no longer the case. The Arbitration Act of 1950 prevented the appointment of the Engineer as arbitrator and the courts have ruled that the 'quasi-judicial' role does not exist. The disputes procedure now included in the ICE Conditions of Contract (and which is examined in more detail in Chapter 13) allows both contractor and employer to refer the decisions of the Engineer to an independent arbitrator who has the power to investigate the matter and, if necessary, overrule the Engineer, even whilst the works are in progress.

This may seen to undermine the position of the Engineer, but through his still-considerable authority, and in particular his function as the initial referee in any dispute, he remains a pivotal figure in the administration of the contract. Whatever the courts have ruled about his role in the eyes of the law, Engineers must continue to exercise their powers in a 'quasi-judicial' frame of mind if they are to maintain that vital position.

Chapter 3
Running the contract

The Engineer's representative: the resident engineer

The Engineer for the works is most unlikely to be able to give his undivided attention to the supervision of the contract for he will have other duties, other contracts to administer, perhaps, and will have only limited time to spare for the main aspects of its management, let alone the detailed control of the work on the site. The contract recognises this situation and allows for the appointment of an Engineer's representative 'to watch and supervise' the construction process.

The title reflects the legal rather than the operational position of the holder of this post, and is very much a product of the draftsman of the ICE Conditions of Contract; in the GC/Works/1 form the term 'Supervisory Officer' is used. The more descriptive designation of 'resident engineer' is as old as the profession itself (it was in common use in 1826 when Marc Brunel, Engineer for the Rotherhithe Tunnel, appointed his son Isambard to be in charge of site supervision), although as far as the ICE Conditions of Contract are concerned it is a courtesy title only. Nevertheless, to most of the staff on a construction project the head of the supervisory team will always be 'the R.E.' and in describing practical aspects of site supervision that name will be used in the following pages; but, for what the lawyers call avoidance of doubt, the proper contractual title will continue to appear in quotations from the ICE Conditions of Contract and any direct comment upon them.

The basic powers of the resident engineer are closely restricted and are in fact almost wholly negative in nature: his task is to ensure the contractor constructs the works in accordance with the specification, drawings and other contract documents and consequently he can only instruct the removal, replacement or re-execution of improper materials or workmanship. He has no inherent power to order additional or varied work, or to approve anything which goes outside the detailed requirements of the contract documents. Indeed, the only positive power allowed him under the contract, which he shares with the Engineer, is that of appointing his assistants (i.e. the site team) whose authority is similarly limited to enforcing the provisions of the specification and drawings.

In practice, very few sites operate within such narrow constraints. The delays and consequential costs arising out of continuous referral back to the Engineer would be prohibitive, and the absence of any detailed personal knowledge of the progress and problems of the works would in any case force the Engineer to rely heavily on the recommendations of his representative. The natural result of this situation is that the resident engineer is effectively the decision-maker as far as a large part of the Engineer's duties is concerned: either he short-circuits the procedures in those areas and acts as if he were the Engineer, or the Engineer formally delegates the appropriate authority to the site. The latter course enjoys the benefit not only of prudence but of legality as well.

Delegation of powers

The contract allows the Engineer to delegate considerable powers to his representative provided the contractor is notified beforehand. However, certain of the Engineer's powers are specifically excluded:

(1) assessment of delay and extra cost arising out of unforeseen physical conditions or artificial obstructions (clause 12(3));
(2) assessment of any extension to the contract period (clause 44);
(3) issue of the certificate of completion accepting the works as substantially complete (clause 48);

(4) issue of the certificate setting out the amount finally due under the contract (clause 60(3));
(5) issue of the maintenance certificate in respect of the satisfactory construction, completion and maintenance of the works (clause 61);
(6) certification to the employer of the contractor's substantial default necessitating forfeiture of the contract (clause 63);
(7) formal adjudication in disputes between the employer and the contractor (clause 66).

The exercise of these functions is considered to be of fundamental importance to the contract and is therefore kept within the direct responsibility of the Engineer who must act in each case after a proper and personal assessment of all the circumstances and information. He may, of course, take advice and receive recommendations from the site staff, but it must be the Engineer's own judgement which determines the result and he will be liable for exercising reasonable professional skill and care in his actions.

The criterion that proper and personal attention be given to every aspect of the Engineer's duties is a good basis for making decisions on what is to be delegated. If the Engineer considers that he will not be able to give the required attention to any part of his delegable functions, then he should pass them on to someone who can. A prime candidate will be the resident engineer, whose proximity to, and involvement in, the day-to-day running of the contract places him in an excellent position to give proper and personal attention to a whole range of matters. Some Engineers draw back from delegating all the permitted powers to the site (and some employers encourage this view), as this means devolving very comprehensive authority to what may be a distant office, served by less than perfect communications and subject to the rush and bustle of site life. This caution is not without its merits, provided it is backed by a reasoned assessment of what to delegate and what to keep back; the best division is one which gives the resident engineer the authority to act in those areas where relatively rapid decisions and/or specific knowledge of the site is needed, leaving issues of longer range and with wider consequences for head office.

A typical set of delegated powers for the resident engineer would comprise:

Clause 13(1), (2)	the test of 'satisfaction' for all work done under the contract, approval of mode and manner of construction;
Clause 14(1)–(6)	approval, revision and investigation of the contractor's programme;
Clause 15(1)	assessing the adequacy of the contractor's superintendence;
Clause 18	instructing boreholes or exploratory investigations;
Clause 19(1)	assessing the adequacy of measures for the safety and security of the site;
Clause 20(2)	instructing repairs to the works during the construction period;
Clause 31(1)	instructing the provision of facilities for other contractors employed on the site;
Clause 32	instructing on how to deal with fossils, antiquities, etc.;
Clause 33	approval of the clearance of the site on completion;
Clause 35	instructing the provision of plant and labour records;
Clause 36	approval of materials and workmanship and the right to carry out reasonable testing;
Clause 37	right of access to all parts of the site and any sources of materials, components etc. off-site;
Clause 38	approval of work which is to be covered up; instructing the exposure of any work for inspection;
Clause 39	instructing the removal, re-execution or replacement of improper work and materials;
Clause 45	granting permission for night or Sunday working;

Clause 49	approval of the works at the end of the maintenance period; instructing the making good of defects notified within the maintenance period; assessing the liability for defects;
Clause 50	instructing the execution of tests or trials to determine the cause of defects;
Clause 51	instructing variations (usually with an upper limit applied to the estimated value of any single variation order);
Clause 52(3)	instructing work to be carried out under daywork rates;
Clause 52(4)	instructing the keeping and maintenance of records associated with any notified claim;
Clause 53(5), (6)	instructing the provision of details of ownership of plant, granting permission to remove plant from the site;
Clause 56(1), (3)	measuring and valuing the work done;
clause 58(6), (7)	instructing the execution of work covered by a provisional item or a nominated sub-contract; instructing the submission of details relating to nominated sub-contracts;
Clause 60(1)	receiving the monthly statement of account;
Clause 62	assessing and instructing emergency repairs.

This pattern of delegation leaves certain clear areas of responsibility with head office: approval of sub-contractors; action on unforeseen physical conditions or artificial obstructions; the power of suspension and the authority to assess the value of any delay, extra cost, change in quantities and ordered variations.

In all these matters careful and deliberate consideration is necessary, drawing on as wide a range of experience and information as possible, and head office may well be a more appropriate place for the decision than the site, depending upon the expertise available.

Two important points need to be made. First, if it is decided to delegate only partial authority to site level, arrangements must be made to ensure that the 'residual' powers (i.e. those delegable powers not passed down to the resident engineer) do not fall into some vacuum of unspecified authority amongst the senior staff at head office. If the Engineer cannot exercise these powers himself, should he make a formal delegation to a suitable member of his staff who is able to keep in touch with the site closely enough to give them the proper degree of attention? Clause 2 of the ICE Conditions of Contract permits delegation to 'the Engineer's Representative or any other person responsible to the Engineer . . . either generally in respect of the contract or specifically in respect of particular Clauses'. It is not clear whether the choice offered by the ICE Conditions of Contract is an exclusive one: that is, between delegation either to an engineer's representative or to a 'person responsible to the Engineer' but not to both. Certainly 'dual delegation', with the resident engineer and a member of the head office staff exercising overlapping authority, is a recipe for disaster and can be eliminated on common sense grounds if not for its legal failings. 'Three-tier' delegation, however, is not open to the same objections provided all concerned are clear on the position: the Engineer holds the non-delegable powers, the resident engineer exercises partial, and specific, authority and a 'person responsible to the Engineer' carries the remaining balance of the delegable powers. It would seem sensible, therefore, to interpret the ICE Conditions of Contract in this way and thus give formal recognition to a working arrangement rather than pretend that the strictly contractual approach is being followed, when so often it manifestly is not.

The second point concerns the use of delegated powers. Any system of delegation can only work effectively if it is operated strictly in accordance with the arrangement notified to the contractor. If the site does not have the authority for, say, the approval of sub-letting then approval should come under the Engineer's signature or, if the three-tier system is in operation, that of the member of the Engineer's staff who holds the necessary authority. The only possible alternative which can be adopted in the interests of streamlining communications is for the resident engineer to use a form of words which make it clear that he is passing on a decision which he cannot make himself, for example: '. . . I am instructed by the Engineer to advise you of his decision as follows . . .'. The contractor can then seek confirmation if he is in any doubt and the resident engineer cannot be accused of giving the false impression that he is allowed to exercise powers beyond those notified to the contractor. The device is nevertheless a dubious one, and open to abuse. It should be used with care, backed always with a proper procedure for obtaining and recording the instructions and decisions involved.

Whatever the extent or method of delegation, contractors are usually allowed the right of appeal to the Engineer. The ICE Conditions of Contract provide that any decision taken under delegated powers can be referred back for confirmation, cancellation or variation; the procedure is kept separate (in clause 2) from the main disputes procedure (clause 66) and so allows an opportunity for second thoughts before the full machinery of a formal Engineer's decision is invoked. In the same way the contractor can ask the resident engineer to examine any instruction given by one of his assistants, thus providing a dissatisfied contractor with a low-key first stage in settling any differences.

The employer has no comparable facility. If he wishes to question any action of the Engineer or his staff he must go directly to a formal decision under the disputes procedure, the final outcome of which could be the embarassing situation of the Engineer and employer facing each other before an arbitrator. This provision emphasises what should in any case be obvious: the need for staff all the way up the chain of delegation to keep their supervisors fully informed, and for the Engineer in his turn to report contentious matters to the employer as they develop.

The contractor's agent

Clause 15 of the ICE Conditions of Contract requires the contractor to superintend the construction of the works either by being present in person (an unlikely eventuality, given the nature of modern civil engineering companies) or by appointing 'a competent and authorised agent or representative' to act on his behalf. It is interesting that the use of the phrase 'agent *or* representative' means that this person need not be, in strictly legal terms, the agent of the contractor with authority to enter into binding agreements or undertakings: he is, nevertheless, universally known as 'the agent' and with the resident engineer makes up the pair of dominant personalities on any construction site.

Whereas the ICE Conditions of Contract are generous in their detail about the extent of the engineer's representative's powers and the arrangements for delegation, they are almost silent on the matter of the contractor's agent. As already noted he cannot be assumed to be a fully empowered legal agent, but on the other hand he must be 'in full charge of the Works'. What does this mean? In practice, the agent, like the resident engineer, is subject to control through his superiors at head office acting directly or through the internal procedures of the firm: there may be a ceiling on his authority to commit the company financially; he may have to submit agreements with sub-contractors for approval; he may have to hire plant through a central department and so on. The Engineer (or his representative) cannot be concerned with these arrangements, which will vary from firm to firm and may exist only in the form of 'unwritten law' rather than as definite policy. He must be able to rely on the prompt execution of instructions without delays arising out of extended chains of command, and the requirement for a contractor's agent 'in full charge' entitles him to just that.

The agent's freedom from direct control can be gauged by the frequency with which the site is visited by head office staff, usually in the form of the contracts manager (this is not an infallible guide as several other factors may be involved, ranging from the financial performance of the contract to its location). Contracts managers and directors of construction companies are not in any way equivalents of the Engineer, but although the agent has 'full charge' it is generally recognised that there is a level of authority above that of the site where important matters arising out of the progress of the works may need to be considered. In some cases this may be the re-examination of a decision or action of the agent following an approach from the Engineer — a similar, albeit informal, version of the contractor's right to have site decisions referred back to the Engineer; on other occasions the agent himself may wish to have, or be told to await, instructions from head office on some sensitive point — a division of authority loosely corresponding to the arrangements for partial delegation of the Engineer's powers. The concept of concentrating full control at site level is subject, therefore, to these practical limitations which are broadly understood throughout the industry without ever being precisely defined.

A set of additional drawings was issued by the Engineer on a large drainage contract accompanied by a clear statement that they were merely a clarification of the specification and not a variation. The decision to give notice of a major claim for extra payment was passed up from the site to the firm's board of directors. The seriousness of the dispute, and its possible effect on the reputation of the company, were considered sufficient grounds for such action. Consequently, the inevitable delay in submission did not disqualify the claim because the whole process was not unreasonably prolonged. (This is the case of Tersons *v* Stevenage Development Corporation (1965) Q.B. 37)

The organisation chart of a typical construction company is shown in Figure 1. Many firms are part of larger groups, often under the umbrella of a 'holding company' which may have varied interests in other fields such as shipping or mining. Within the group, and perhaps sharing a similar name with the construction firm, there may be 'sister companies' operating in such related areas as ready-mix concrete and aggregate extraction. Invariably these undertakings are independent cost centres and it is quite wrong to assume they are in a position to offer preferential treatment or friendly co-operation to an associated company. The surfacing on a motorway contract may be supplied and laid by a firm in the same group as the main contractor, but the relationship will be strictly a business one and subject to all the usual commercial pressures.

The agent's responsibility is to ensure, as far as site operations are concerned, that the contractor's obligations are discharged. This involves:

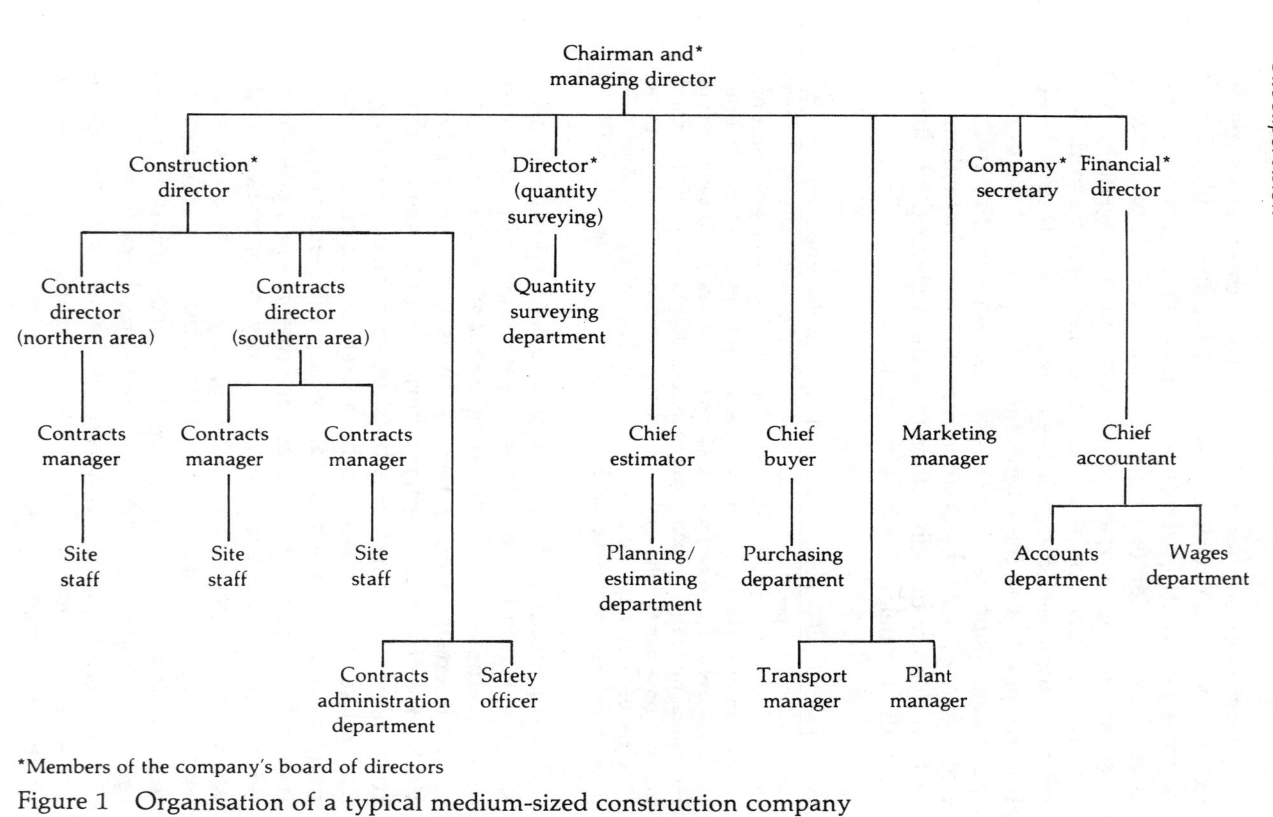

*Members of the company's board of directors

Figure 1 Organisation of a typical medium-sized construction company

(1) constructing, completing and maintaining the works within the contract period, providing all the required labour, materials, plant etc. and avoiding any unreasonable noise, disturbance and damage;
(2) carrying out the instructions of the Engineer and his representative;
(3) providing adequate superintendence;
(4) setting out the works;
(5) providing (and, when requested, revising) a programme together with descriptions of methods of working;
(6) undertaking the care of the works (including temporary works) during the construction period and making good any damage;
(7) providing for the safety and security of the site and all construction operations;
(8) drawing up and submitting regular returns of labour and plant;
(9) preparing and presenting monthly statements of the value of work executed under the contract;
(10) notifying any claims for additional payment, supported by such contemporary records and particulars as are necessary for proper investigation and assessment.

The agent usually has at least one sub-agent to act as his deputy and assist in the management of the site and its labour force. Sub-agents are equivalent to the more senior assistants on the resident engineer's staff and may be given considerable authority about the works; they do not share the same contractual status, however, as the ICE Conditions of Contract make no reference to such an appointment and consequently give no indication of the powers which a sub-agent might exercise. In view of the very limited authority granted to the assistants, this is not a serious disadvantage and may, indeed, be positively beneficial from the contractor's standpoint as it affords some very useful freedom of manoeuvre.

One factor often forgotten by the resident engineer's staff is that the agent must operate in a commercial environment. Reputable construction companies are as concerned about engineering standards as any designer or supervisor, but they are in business to make a reasonable profit and whatever the level of individual or short-term losses which can be sustained, no firm can continue trading at a loss indefinitely. Agents are consequently under pressure to show a realistic rate of return, for it is on this that the company's future, and their own prospects, must rise or fall. Accordingly, contractors cannot base their decisions solely on abstract engineering criteria and supervisory staff would do well to take a practical, and understanding, view of the very real problems of achieving standards and meeting deadlines at a profit.

Take for instance the supply of ready-mixed concrete. Large regional depots can offer the prospect of fast turn-round and plenty of reserve capacity but in reality the contractor may have to take his place in a queue of several large customers and may find the plant unresponsive to the needs of individual sites. A small local firm with less impressive facilities and a greater risk of interruption in supply may nevertheless be more willing to accommodate the requirements of a substantial and valued client. The agent will balance these factors, plus the size of the discount he can negotiate, and may conclude that a deal with the smaller plant is a calculated and justifiable commercial risk; contrast this with the resident engineer's staff who, basing their judgement on 'pure' engineering considerations, might see it as an ill-informed gamble on an unreliable and second-rate supplier.

As another example, consider two local sources of graded stone fill. If one is 20 per cent cheaper than the other, the agent can afford to accept a rejection rate of, say, 10 per cent and still, other things being equal and the disruption effect kept under control, show an advantage over the second source even if it is of uniformly high quality. The resident engineer's staff may feel the loss of one load in ten to be clear proof of a slapdash and ill-managed approach, but the agent will be well satisfied, especially if he has another destination where less stringent requirements apply to which he can divert his rejected loads.

In both these examples the agent is relying on his arrangements turning out to be 'satisfactory', knowing that they are not the best available. It can be argued that in doing so he is technically in breach of the contract by knowingly offering materials or workmanship which stands a significant chance of rejection, but this is a very academic position to take up and entirely divorced from the realities of the construction industry. Similarly, it can be said that the agent is using the resident engineer's staff for the quality control and superintendence which the contractor has agreed to provide. This is also a view which has some contractual justification, but little else to recommend it as a working principle for site supervision. Of course a point could be reached when the resident engineer (or the Engineer) has to conclude that the agent's commercial instincts are working to the detriment of the engineering standards on which the acceptability of the works depend. At that point, or ideally just before, the provisions of the contract can be used to put matters right, but each instance should be judged on its merits and in the light of experience which recognises the practical side of civil engineering contracting.

Decisions like these are often difficult and can sometimes appear as dubious compromises, particularly to staff who are not kept fully informed of the issues involved, but if all site decisions were easy the resident engineer and agent would lead very dull lives indeed!

Chapter 4
The contract documents

One of the most notable features of modern construction projects is the sheer volume of documentation. Even for minor schemes such as a small pumping station or a roundabout the drawings, specifications, schedules and bills of quantities will be as much as a man can carry. A relatively modest section of motorway with a dozen or so structures generates about 300 A1 size working drawings and layouts, an A3 folder depicting a further 70 or 80 typical details and standard arrangements, six or seven 200-page volumes of contractual requirements and data (including photo-reduced computer printouts), a bill of quantities running to well over 400 pages plus a similar quantity of associated schedules, two thick and close-printed volumes covering the standard specification and method of measurement and a small library of codes of practice, British Standards and so on.

There is little evidence to suggest that the trend towards complex and very detailed documentation has effected a reduction in costs or the incidence of disputes and many would say that the result has been quite the reverse. In fairness, however, some weight must be given to the counter-argument which explains this development as a reaction to ever larger and more technical claims. It is not clear whether the instigation has come from contractors, Engineers or employers (through their ancillary professional advisers, the quantity surveyors, auditors, etc.), but what does seem clear is the strong and paradoxical view across the industry that the trend should be deplored and has to be accepted as permanent.

It was not always so, and once again a glance back to the 'golden age' of engineering reveals an illuminating example to contrast with the present situation. George Stephenson, as Engineer for the Liverpool and Manchester Railway Company, came before a House of Commons Committee in May 1825 to be examined on the accuracy of the designs and estimates which he had prepared in support of the Parliamentary Bill necessary for this undertaking. Edward Alderson QC appeared for the opponents of the railway project and he was interested in details of the proposed engineering works, particularly the crossing of the River Irwell.

Alderson: *What is the width of the Irwell here?*
Stephenson: *I cannot say exactly at present.*
Alderson: *How many arches is your bridge to have?*
Stephenson: *It is not determined upon.*
Alderson: *How could you make an estimate for it then?*
Stephenson: *I have given a sufficient sum for it.*

Stephenson would presumably have refined his pre-contract planning somewhat had the Bill been passed (it was not, and a second attempt had to be made a year later), but this example demonstrates how sketchy the preparation work on major engineering schemes might be. Estimates generally covered only 40 to 50 per cent of the final cost and Engineers invariably did not bother to furnish tenderers with detailed specifications or accurate quantities and consequently the contract sum was likely to be as wild a guess as the Engineer's own estimate.

Joseph Locke, who with Brunel and Robert Stephenson formed the 'great triumvirate' of Victorian engineering, developed the system of detailed contract documentation allied to close site supervision and measurement. Less well known today than his two great contemporaries, Locke is nevertheless a very recognisable figure to modern civil engineers: a careful, methodical professional with an enviable reputation for getting work done on time and within the estimated cost. His success owed much to the skill and care which went into the preparation of the contract documents, and in that respect nothing has changed over the last century and a quarter. It is also just as true now as in the 1850s that careful preparation is not enough in itself, the contract documents must be used properly and that means they must be understood.

The main components of a typical set of contract documents are: the form of tender and agreement; the conditions of contract; the drawings; the specification; the priced bill of quantities together with any associated schedules. Each part has its own purpose and identity, and separate existence too, as the documents are very rarely bound together in a single volume, but they are mutually dependent and must form a complete description of the contract without contradiction or ambiguity.

Form of tender and agreement

In most tender documents a pro-forma letter is included, addressed to the employer and requiring signature by the contractor, which promises to 'construct and complete the whole of the Works . . . for such sum as may be ascertained in accordance with the Conditions of Contract' (see Figure 2). Usually the form of tender and agreement refers to an attached appendix (see Figure 3) which sets out such essential details as:

(1) the value of the bond which the contractor may be required to arrange through a bank or insurance company to guarantee performance of the contract and which would be forfeit to the employer in the event of the contractor's failure;
(2) the time for completion of the works;
(3) the liquidated damages payable to the employer for each day or week of delay in completing ('liquidated' means that the likely loss to the employer has been ascertained and expressed in money terms);
(4) the length of time for which the contractor must maintain the works after completion;
(5) the minimum value of work which must be completed in any month before the employer will accept an account for payment.

As most forms of tender state that the form itself plus the

All Permanent and Temporary Works in connection with A329 Amen Corner to Downshire Way

FORM OF TENDER

(Note: The Appendix and Schedule form part of the Tender)

TO: The Berkshire County Council

Gentlemen

Having examined the Drawings, Conditions of Contract, Specification and Bill of Quantities for the construction of the above-mentioned Works (and the matters set out in the Appendix hereto), we offer to construct and complete the whole of the said Works and maintain the Permanent Works in conformity with the said Drawings, Conditions of Contract, Specification, Bill of Quantities and Schedules for such sum as may be ascertained in accordance with the said Conditions of Contract.

We undertake to complete and deliver the whole of the Permanent Works comprised in the Contract within the time stated in the Appendix hereto.

Our tender has been based on the supply of materials and manufactured goods as indicated in the Schedule hereto.

If our tender is accepted we will, when required, provide two good and sufficient sureties or obtain the guarantee of a Bank or Insurance Company (to be approved in either case by you) to be jointly and severally bound with us in a sum equal to the percentage of the Tender Total as defined in the said Conditions of Contract for the due performance of the Contract under the terms of a Bond in the form annexed to the Conditions of Contract.

Unless and until a formal Agreement is prepared and executed this Tender, together with your written acceptance thereof, shall constitute a binding Contract between us.

We understand that you are not bound to accept the lowest or any tender you may receive.

We are, Gentlemen

Yours faithfully

Signature FRENCH KIER CONSTRUCTION LTD. Date 7 OCTOBER 1980

Address TEMPSFORD HALL, SANDY, BEDS.

Figure 2 Typical 'Form of Tender'

APPENDIX

Note: Relevant Clause numbers are shown after the description

Amount of Bond	10	:	10% of Tender Total
Minimum Amount of Insurance 23(2)		:	£1,500,000
Time for Completion 43		:	104 weeks
Liquidated Damages for Delay 47(1)		:	£1,500 per day
Period of Maintenance 49(1)		:	52 weeks
Vesting of Materials not on site 54(1) and 60(1)		:	None
Method of Measurement adopted in preparation of Bills of Quantities 57		:	Method of Measurement for Road and Bridge Works (1977) as modified and extended by supplement No 1 dated 1978, together with amendments as described in the Bill of Quantities
* Percentage for adjustment of prime Cost Sums	59A(2)(b) and (5)(c)	:	.....5.....%
Percentage of Value of Goods and Materials to be included in Interim Certificate	60(2)(b)	:	97%
Minimum Amount of Interim Certificates	60(2)	:	£100,000

* The percentage must be inserted by the Tenderer

Figure 3 Typical 'Appendix' to the 'Form of Tender'

employer's letter of acceptance constitutes a binding contract 'unless and until a formal Agreement is prepared and executed', there is no need in law to do more. However, many employers prefer to have an agreement setting out the names of the parties to the contract and summarising their obligations, and in the case of public authorities this is often a requirement of their official procedures. The agreement is made under the seals and signatures of both parties, an ancient practice meant to emphasise the importance of the undertaking, and whilst a more impressive document than the form of tender it adds nothing to the strength of the contract, except that the period in which a legal action can be brought for breach of the contract is extended to twelve years from the date of the breach instead of six.

Conditions of contract

In most cases one of the standard forms of contract (ICE, JCT or GC/Works/1) is used, and these are so familiar and readily accessible that their provisions can be embodied through a brief statement in the contract documents giving the proper title together with any other identifying data such as edition number, date of publication etc. Hence, there is no need to reproduce all the clauses of the ICE Conditions of Contract (or any other standard form) in the contract documents, nor to bind in a copy (although every site will have a set for reference purposes), but it is essential to set out any additions or modifications to the conditions clearly and prominently. Whilst modifications are comparatively rare, additions are common and usually take the form of a series of special conditions introduced by, and following on from, clause 72. The additions most frequently included are contract price fluctuation (cpf) clauses, governing the application of index-linked variations to labour and materials costs, and special requirements clauses, covering the particular provisions relating to statutory bodies such as British Rail and the Central Electricity Generating Board, whose operations or apparatus may be affected by the construction of the works.

These additional clauses will normally be standard versions drafted for general application and not tailored to the specific needs of a particular project, and the contractor will probably have worked with them, or something similar, before. Nevertheless, special requirements clauses in particular are a potential problem area as many of them contain a very large number of provisions, and some aspire to be both broadly comprehensive and minutely detailed at the same time. Furthermore, it is not unusual for these provisions to grant rights and powers to officials who are mentioned nowhere else in the contract. For instance, where the works affect British Rail operations the standard special requirements clause gives the divisional civil engineer certain powers of approval and intervention independent of those exercised by the Engineer for the works. There is, therefore, every reason to ensure that all concerned are made aware of any additions or modifications to the basic conditions of contract.

It is important to remember that supplementary powers or obligations introduced through additional clauses are never intended to supplant those still contained in the basic conditions of contract — they are to run in parallel. So compliance with the provisions of a special requirements clause does not relieve the contractor of any of his other liabilities, no more than the fulfilment of his obligations under the basic conditions of contract could leave the contractor free to ignore the special requirements.

On a contract where deep excavations exposed a high pressure water main, the Water Board approved the contractor's proposals for supporting their main but the Engineer insisted on additional temporary works. The contractor had to satisfy both the Water Board (under the provisions of an additional special requirements clause) and the Engineer (through the normal application of the conditions of contract) and thus there was no restriction on the Engineer's power to instruct any necessary and reasonable measures to protect the works.

On the other hand, when a contractor had to drive piles adjacent to a railway bridge the Engineer's approval of the proposed method of working did not overrule British Rail's insistence on a different, slower method which produced less vibration. The original proposal satisfied the Engineer's requirements as far as the construction of the permanent works was concerned but the provisions of a special requirements clause gave British Rail the power to demand safeguards in respect of their bridge.

The drawings

The standardisation of the conditions of contract, the specification and the bill of quantities has advanced so far in recent years that many engineers see the drawings as the only part of the contract documents where some individuality can be expressed and real professional judgement exercised. For this reason, and because one of the engineer's basic skills is the translation of design concepts into construction plans, the drawings are generally considered to be the heart of the contract documents, the single most important component.

The drawings comprise two basic types:

(1) layout and arrangement drawings covering particular aspects or parts of the works at various scales depending upon what information is being conveyed; thus the whole site may be included at 1:2500 scale to show public rights of way and permitted means of access, or a single prestressed concrete beam may be drawn at 1:50 scale to show the reinforcement pattern.
(2) typical details depicting standard components which will be repeated throughout the works and which may be represented only by a symbol or code letter on the layout drawings; examples are pre-cast concrete manholes, steel guardrails and lighting units.

The contract documents must contain a list of all the drawings, each with a brief explanatory title and unique number. Frequently a suffix is added to the drawing numbers in order to emphasise the special status of the set issued for the contract, 'C' for contract drawings, or 'W' for works drawings.

The drawings will incorporate any amendments or additions notified during the tender period, all of which are considered to be covered in the contractor's offer. Any subsequent changes or additional drawings may constitute variations to the contract involving the contractor in extra cost or delays for which he will be able to seek payment. New drawings, or new versions of drawings already issued, will be numbered in accordance with the system used for the contract drawings and recorded in a register kept on the site.

It is worth noting that the Engineer has power under clause 7(1) of the ICE Conditions of Contract to issue 'from time to time during the progress of the Works such modified or further drawings' as he decides are necessary. On the other hand the contractor, under clause 7(2), must give 'adequate notice in writing' of any further drawings which he might consider are required. Taking a strictly literal interpretation of these provisions it could be argued that the contractor has no remedy for late drawings if they are not formally requested, and that modified or additional drawings can be issued without any liability on the part of the employer provided their arrival is not so timed as to actually disrupt the execution of the works. Anyone who sets out to prepare a set of drawings, or to supervise their use by a contractor, with this idea in mind is looking for trouble and will deserve it when it appears (as it undoubtedly will). Of course, there will be occasions when an additional drawing, although late, will be issued in time to avoid abortive work or disruption and without invalidating the assumptions on which the contractor tendered, but more often than not there will be some consequential effects. It is hard to imagine how any modification to an existing drawing, unless trivial, can fail to upset the contractor's tender assumptions, and hence his estimating, to some degree.

Site staff must accept that no set of drawings is ever perfect and should adopt a practical approach with the following objectives in mind:

(1) minimise alterations and additions;
(2) keep those that are essential as simple as possible;
(3) issue them as soon as practicable;
(4) identify promptly (and with the contractor if possible) any consequential costs and delays.

During the tender stage on a motorway contract an amendment was issued changing the type of bearings to be used on a bridge. Appropriate alterations were made to the specification and bill of quantities and the contractor was able to price the revised bearings in his offer. After work had begun the contractor asked for details on how to mount the bearings as the drawings showed the arrangement for the original type. An additional drawing was prepared and issued giving the required information and also showing some minor changes to the dimensions of the bearing plinths. Although the drawing was 'late', construction had not proceeded beyond foundation stage at the bridge and so no abortive work was involved.

The modifications to the plinths involved no extra materials and had only a negligible effect on any estimated labour element. No consequential costs or delays could be identified.

A factory being constructed on a new town industrial estate had steel roof beams. A late drawing was issued clarifying the method of mounting certain service ducts alongside the beams and requiring some additional welding to fit support brackets. The ducts had not been fitted and so abortive work was largely avoided. The beams, however, were all in place and although the contractor had intended to carry out all steel fabrication in his workshops, the extra welding had to be done on site with men working off temporary access platforms rigged in the roof. Thus, not only was there additional work, but it had to be done under conditions not envisaged at tender stage and which significantly reduced output on that operation as well as disrupting adjacent activities. The contractor suffered direct and consequential losses which he was entitled to recover.

The drawings are nowadays frequently supplemented by reams of computer printout containing setting-out and alignment details together with other information such as earthworks volumes. The use of computer techniques, particularly three-dimensional modelling, has greatly assisted the design process and results in a great deal of the data which formerly appeared on drawings now being presented in purely numerical form. The computer has revolutionised many aspects of civil engineering, and site staff, both in the agent's and the resident engineer's offices, have every reason to be grateful for the way it has taken much of the drudgery out of routine setting-out and measurement calculation. The benefits are clear, provided the maxim of 'rubbish in — rubbish out' is given due regard. Printouts are accurate representations of the computer's calculations, which is not to say that they are accurate contractually. Incorrect or inaccurate basic data will produce an incorrect or inaccurate result from a computer just as surely as it will on a drawing board, except the former may be more difficult to detect.

In general, an engineer will recognise an error on a drawing more readily than one in a long series of almost identical figures. Carriageway alignments, earthworks sections and setting-out traverses are all prime candidates for computerisation as they involve very large numbers of calculations, and often tedious re-calculation, to obtain the desired end-product. They are also, once calculated, easily represented graphically or diagramatically, thus conveying their information with clarity and also focusing attention directly on any errors. This kind of data, when presented as print-out, should also be depicted on the drawings in the 'old fashioned' way, and if not, the resident engineer's staff should spend the relatively quiet period at the very beginning of the contract transferring it from the printouts to drawing paper.

Drawings are a very effective means of communicating information, although not without their limitations. The practice of adding notes to contract drawings should be treated with special caution. Often, to avoid cluttering up the detail of a drawing with close-printed lettering, a series of explanatory notes may be placed in a separate panel. Sometimes this device is taken much further and the contents of these notes go beyond mere explanation to cover such matters as approved methods of working, restrictions on the choice of types of material or suppliers' components, performance requirements and so on. Matters, indeed, which are normally dealt with in the specification. It is true to say that no less weight should be attached to a note included in the drawings than to a clause in the specification and the contractor is obliged to take account of both. However, there is a better chance of avoiding ambiguities and contradictions if the function of such notes is confined to explanation and any other objectives are achieved through the specification. The drawings and specification should complement each other, not act as substitutes. It is of no benefit to the contract if the contractor misses an essential requirement because he did not expect to see such important matters contained in a small note on one of the drawings. The Engineer may decide that the employer is entitled to performance of the requirement at no extra cost, but the contractor's loss, especially if substantial, will inevitably have its consequences on the conduct and quality of the remaining operations.

To stress the need for clarity in the drawings is perhaps to state the obvious, but what appears perfectly plain to the designers and their colleagues on the supervisory staff may be ambiguous or downright obscure to a contractor. Well-presented, clear drawings can convey the contractor's obligations better than whole clauses of the conditions of contract and pages of the specification, and may prove to be the fundamental element in any dispute over implied terms. The best test of the adequacy of the drawings is to sit before them and, adopting the point of view of a contractor coming fresh to the information, attempt honestly to answer the question, 'how can this be built?'

Accuracy is the other obvious yet essential requirement, not only in depicting what is to be constructed but also what already exists on the site. In the eyes of the law an error, even an innocent one, can amount to misrepresentation.

The specification

The specification amplifies the contractor's duties and obligations as set out in the conditions of contract, giving detail on such matters as methods of working, types of plant, characteristics of materials, and so on. A properly drafted specification does not conflict with the conditions of contract, nor free the contractor from any of its requirements.

The specification is rarely, if ever, exhaustive: specific details are not set down in respect of every operation, and where such details are given, they are not always complete and definitive. A typical specification contains a number of clauses affirming certain positive standards which must be achieved (concrete strengths, for instance), together with a corresponding set of clauses listing other features which must be avoided (it may, for example, impose a ban on the use of frozen aggregates). These 'shall' and 'shall not' clauses will not cover all aspects of the contract and the gap is filled by a third type of clause which provides guidance for the contractor on the means by which he can discharge his obligations satisfactorily. It usually does this by setting down options, ranges, tolerances and similar data (for example, a permitted grading 'envelope' for carriageway sub-base).

The inter-relationship of specification clauses with each other and with the requirements of the contract conditions is of vital importance to the contract, and the resident engineer (and sometimes the Engineer himself) will be involved in many decisions of interpretation. Within the legal requirements of the contract and the technical requirements of the permanent works there is often room for flexibility in solving problems of this kind, with two provisos:

(1) when the specification offers some degree of choice, the contractor must accept the consequences and their subse-

quent effect on other operations (of which more will be said in Chapter 9);

(2) when the specification sets down fixed requirements, the contractor should never plan for anything less than rigid adherence to them (the oft-repeated complaint of 'over-zealous application of the specification' is a contradiction in terms).

As with so much modern contract documentation, the specification is often standardised and designed for general use. Most of the large promoting organisations with a continuing programme of work, such as British Rail, the various Water Authorities and the Property Services Agency, have their own specifications. Amongst the best known is the Department of Transport's 'Specification for Road and Bridge Works' which is used on almost all projects of this kind in the United Kingdom whether carried out for national or local government departments, public sector employers or private developers. These standard specifications, like the ICE Conditions of Contract, are printed and issued separately and are so well known that they are incorporated into the contract documents by reference only. This independence and familiarity allows them to develop an identity of their own: the 1976 edition of the Department of Transport specification is treated as something of a design aid as well as a contractual document and is universally referred to as 'the Blue Book' from the colour of its cover (thus distinguishing it from its predecessor which was known as 'the Orange Book' or, less respectfully, 'the Orange Peril').

On almost every contract where a model form of specification is used there will be special features not properly covered in the standard document which require the addition of new clauses or the modification of existing ones. Tenderers will generally take such a specification at its face value (one of the advantages of standardisation is faster and more certain estimating) and so it is essential to draw their attention to any changes by listing them prominently.

The usual convention is to assign the original clause number plus the suffix 's' (for 'substitute'), to amended clauses and the suffix 'c' to cancelled clauses. New clauses are given numbers which run consecutively from the last original clause in their appropriate group or chapter. Great care must be taken to amend cross-references to these clauses throughout the whole specification and to check for contradictions. The committees that draft the standard specifications frown upon alteration or piecemeal revision of their efforts and it is common practice in advice to users to encourage as an alternative the incorporation of notes on the drawings, a device which has been criticised earlier. Consistency has many worthwhile benefits, provided it is not made an end in itself. The resistance of draftsmen to changes is understandable, if not always founded on practical considerations, but then few drafting committees have to administer their specifications on site.

At the drafting stage two models or prototypes can be used:

(1) the performance or 'end-result' specification in which the standard of the end product is defined in terms of its final strength, durability, functional properties, etc. The contractor is expected to take into account all the risks associated with guaranteeing that his chosen methods and materials can produce such a result.
(2) the method or 'recipe' specification which provides the contractor with detailed descriptions of how the various components of the works are to be produced and so places the risk for the successful performance of the finished product with the designer and, ultimately, the employer.

The object of both methods is to bring certainty of cost to the contract, in the first case by transferring away all the unknown quantities and in the second by designing them out. Arguments are advanced to the support of each of these approaches as the best means of ensuring cheap contracts. The performance specification allows the contractor the opportunity to use his ingenuity and experience to win the work by devising the most economical solutions to construction problems; the method specification sets out clearly defined, quantifiable requirements on which estimates can be based without the need to build in extravagent contingency allowances.

Few specifications in the United Kingdom are based exclusively on one or other of these prototypes. Most are drawn up using a combination of both, the Department of

Transport's specification being a typical example where the performance approach in clause 1610, Compaction of Concrete, 'All concrete shall be compacted to produce a dense homogeneous mass . . .', can be contrasted with clause 608, Compaction of Embankments and Other Areas of Fill, which provides a comprehensive 'recipe' listing approved techniques and types of equipment. Sometimes, individual clauses are drafted with both performance and method requirements included. Again, an example can be found in the Department of Transport's specification. Clause 605, Excavation of Foundation Pits and Trenches, leaves the contractor to decide how to achieve the end-result of 'adequately supported' excavations, but goes on to give details of the method to be used to fill any soft spots found in the base of the dig. Mixing both the performance and method prototypes within a specification, or even within a specification clause, can be worthwhile so long as it is not done at the expense of clarity and provided that performance and method are not used to specify the same requirement, except in those rare cases where the draftsman is certain that the stated result can be achieved with the stated method.

To supervise effectively the resident engineer and his staff must have more than just a comprehensive knowledge of the contents of the specification, they must be aware of the reason behind each requirement and of the complex relationships between requirements, otherwise they will be making decisions in artificial compartments. This important topic will reappear in Chapter 9 when the process of approval and acceptance is considered.

The perfect specification has not been drafted, and all versions are compromises between the need to lay down what is required for the completion of the permanent works and the desire to avoid placing unnecessary and expensive restrictions on contractor's flexibility. The risks of vague or incomplete specification are self-evident, but for those having to enforce the requirements on site there are real disadvantages in the corresponding vice of 'over-specification'.

Consider the problems associated with a specification clause covering free-draining material. If the performance model is chosen as the framework, then the core of the clause will be a definition of the drainage characteristics which a material must exhibit to be accepted. This might be done by describing a test to measure, say, the porosity of material under standard conditions and specifying the minimum result. Certain practical difficulties arise. The introduction of a laboratory test means representative sampling and the need to take the material to the site laboratory (or wherever else testing is carried out) with the consequent likelihood that results will lag behind deliveries on what is usually a fast-moving operation; bulk free-draining supplies are rarely uniform and results will vary widely and confusingly unless the test has been designed with a low (but not too low!) level of sensitivity. Moreover, the test will not measure other important characteristics such as load bearing capacity, stability under the action of running water, particularly the resistance of the finer material to washing-out, and frost susceptibility. There are many materials available which will pass a drainage test but fail to meet other operational standards and to overcome this problem further details can be given to guide the contractor: recognised poor materials can be specifically excluded or preferred sources listed. This is, of course, the incorporation of a 'recipe' element and the clause could fall into the trap of combining performance and method requirements which are not fully compatible.

The same difficulties appear in different guise if the 'recipe' model is used. Here the technique might be to list the acceptable classes of material and provide a range of size gradings (this is an approach in general use in current specifications). Experienced site staff know materials which satisfy such a test of acceptance but which at the same time, are notorious for their unwillingness to drain or for poor performance under construction traffic, a failing in the 'recipe' which could be remedied by adding a performance requirement such as a drainage test.....

It can be argued that the comprehensive requirement discussed in the example above can be expressed in much simpler words, 'free-draining material shall be suitable fill in accordance with the contract which will drain freely'. The success of such a clause depends entirely on the competence of Engineers to give 'fair and reasonable' judgements on the performance element, the ability of contractors to anticipate what that standard might be and the capacity of both to behave consistently. It is a disappointing feature of modern contract administration that such aims are considered over-optimistic.

The bill of quantities

The bill of quantities is a schedule listing the constituent parts of the works with identifying descriptions, estimated quantities and priced rates for each item. The function of the bill is to provide data for the comparison and assessment of bids at tender stage and thereafter to act as a schedule of rates for valuation purposes and to assist the Engineer in assessing the rate for any new or varied work. Under the ICE Conditions of Contract the contract sum, the 'bottom line' of the bill of quantities, has no binding force because it is derived by summing the product of each rate and its appropriate quantity and the ICE Conditions of Contract specifically state that the quantities are not final but are subject to remeasurement.

Thus, as far as the contract is concerned, errors in extending a rate (i.e. multiplying it by the quantity) or in totalling up the various elements of the bill are of no effect. The rates themselves, however, are binding and even if they contain mistakes the contractor will be held to them unless there are exceptional circumstances. One such exception is the case where the employer realises, or is advised by the Engineer, that there is an error in the rates and nevertheless accepts the tender: here the contractor would be able to apply to the courts to have his mistake rectified. This emphasises the value of a detailed check on bill rates at tender stage. If it reveals any errors, an immediate request should be made to the potential contractor to confirm or withdraw his tender. No opportunity should be given for changing the bid as this would allow one tenderer an advantage over his competitors (indeed, there is a belief in some quarters that 'mistakes' are occasionally made deliberately with the aim of gaining such an advantage) and the confirmed rate cannot be re-adjusted later even though the contractor may face a substantial loss.

On a bridge contract in the late 1970s the bill included an item for a number of special pre-cast concrete units to face part of the abutments. In one of the tenders the extended figure carried forward to the summary was £8,000, which was reasonable, but was clearly not the product of the quantity (100) and the quoted rate (£8). A mistake appeared to have been made, probably because the estimator had done the calculation 'the wrong way round' by deciding on a total price first and then dividing it, incorrectly, by the quanity to get the rate. As the possibility of error was known, the employer invited the tenderer to confirm his rates and the contractor, guessing that he must be in a favourable position to win the contract, did so.

All the 100 units in the bill were paid for at this confirmed but clearly uneconomic rate. During the course of the contract it was found that a further 16 units were required, which the contractor had to supply at the same price as there was no difference in the nature of the extra work to justify a departure from the bill rate.

The bill of quantities is normally introduced by a preamble stating the basic principles on which it has been prepared. As with the other components of the contract documents, standard forms are available and are widely used: the Institution of Civil Engineers' 'Standard Method of Measurement' (CESMM), for example, and the Department of Transport's 'Method of Measurement for Road and Bridge Works' (MMRB). Once again, these well known and separately available model documents can be incorporated into the contract by a simple reference in the preamble, although any additions or amendments must be listed clearly and prominently, as with a standard specification. The preamble may also set down rules for interpreting the priced bill, typically stating that rates will be 'fully inclusive' (i.e. covering delivery charges, waste, storage, temporary works, overheads etc.), that any unpriced items will be deemed to be included elsewhere in the bill and that where a choice of alternative materials is permitted the price will be taken to cover them all.

The first section in the bill itself, the preliminaries, covers the contractor's general obligations, such as providing offices and other site facilities and maintaining diversions. The remaining sections may be divided up on an operational basis (foundations, steelwork, finishings, etc.), or by location (western abutment, eastern abutment, main deck, approach roads etc.) or by a combination of both. The bulk of any section will consist of straightforward measurement items in which an estimated quantity is set out for pricing by the main contractor. However, there may be a number of items specifically designated as 'provisional' or 'prime cost' and which require special consideration. A 'provisional' sum covers work for which accurate quantities cannot be predetermined but which, nevertheless, is very likely to be required and so needs pricing. A typical example might be the excavation and disposal of a layer of peat, detected in varying by some of the boreholes at the site of a mass concrete foundation. The 'provisional' label warns the contractor that the quantity may vary significantly and this factor should be taken into account when setting the rate. To assist the contractor in making the rate a reasonable one, the quantity in the bill, albeit uncertain, should represent the best estimate of what is expected: it would be misleading and unprofessional to put in a notional 'round figure' simply to get a rate without giving proper consideration to the amount of work covered by the item. Similarly, provisional rates should not be abused by extending their coverage into areas of new or substantially changed work which could not have been assessed by the contractor in his pricing.

A contract for the construction of a sewage works in the West Country contained a 'provisional' sum for crushed stone. The quantity was very small and no location was shown on the drawings, clearly the item was intended to cover the contingency of filling local soft areas as they were encountered. The stone was used for this purpose as the works progressed, and although the quantity involved somewhat exceeded the figure in the bill, with a provisional sum this was not unexpected. However, when work began on filling an old watercourse it was found that the fill available on site, and earmarked in the contract documents for this operation, performed poorly under compaction and consequently held up progress. The resident engineer instructed drainage layers to be spread alternately with layers of the fill, the instruction specified the use of crushed stone and the provisional rate was applied in the measurement. The contractor disputed the valuation and claimed that the rate should not be applied. Because the total quantity of stone was considerably greater than that given in the bill and the kind of operation (layering into earthworks rather than dumping into soft areas) was so different as to make the estimating assumptions invalid, the contractor was successful in his claim for a new rate for the stone used in the filling of the watercourse.

'Prime Cost' items relate to specific operations or components which are to be the responsibility of a specialist subcontractor working under the control of the main contractor. Usually the Engineer or the employer nominates the subcontractor and the contractor is not given the opportunity of pricing the item, a figure being inserted in the bill when it is drawn up. As the work will involve the main contractor in providing access, working space, site services and facilities and possibly attendant labour, each prime cost item is normally accompanied by two further items for which bids are invited:

(1) a lump sum to cover 'labours', which includes all work and services provided by the main contractor;
(2) a percentage of the prime cost figure, to cover the main contractor's profit and any other charges.

The pitfalls of nominating sub-contractors will be discussed later, in Chapter 10, but it should be noted that unless it is certain beyond reasonable doubt that the work will proceed in the manner specified in the documents, provisional sums should be used rather than prime cost items.

The sub-totals for each section of the bill are carried to a final summary. On this page the grand total of the priced bill of quantities appears, a figure important in the award of the contract yet of no significance in its valuation. Two other entries may also be found immediately before the grand total: a 'contingency allowance' and a 'balancing (or adjustment) item'. Usually expressed as a percentage of the sum of all the bill sections, the contingency allowance is given a 'provisional' label to indicate that it may not be taken up in full or even at all. Its purpose is to ensure that financial provision exists in the funding of the contract to cover unforeseen extras or variations, and in the modern environment of tight budgetary control it is an increasingly rare sight. The balancing item is the last entry before the grand total and permits adjustment of that figure through the inclusion of a lump sum, either positive or negative, without affecting individual rates. Typically, the balancing item is decided by the senior management of a construction firm after the estimators have done their work: a positive adjustment may be made where it is felt that margins have been cut to an uneconomic level or a negative figure may be inserted to produce a very competitive bid. The balancing item is incorporated into the monthly accounts by an appropriate adjustment to the value of the work executed. Negative balancing items can have a profound effect on a contract because they represent the calculated risk that rapid progress, favourable cash-flow, efficient working or just plain luck will produce the corresponding saving on the original estimate, and therefore any miscalculation, mismanagement or misfortune will mean a virtual guarantee of a loss on each month's operations.

The body of each section of the bill of quantities comprises a schedule of numbered or coded items describing the various components of the works. The standard methods of measurement all provide a library of item descriptions, each of which is subdivided in such a way as to permit many different combinations covering a wide range of work. This type of system has the advantage of reducing the preparation of bill items to a step-by-step process which is both quick and consistent: staff time is saved, a degree of uniformity is imposed and the use of standardised elements to build up items makes computerisation possible. Occasionally it is necessary to include a special description for a part of the works not covered by the standard method of measurement, but these 'rogue items' must follow the same format as the rest of the bill and should be used only when strictly necessary.

Each of the standard methods of measurement has its own basic principles and instructions for use and as these are set out in the various documents, no detailed review will be attempted here. The general approach to coverage is relevant, however, and worth consideration. Normally, all contractor's rates are taken as 'fully inclusive' and for this reason some standard methods of measurement (the CESMM for example) set out to avoid any specific description of the tasks to be carried out under any item and concentrate on identifying the components. This approach is not universally adopted and an example of the alternative method can be seen in the 1977 edition of the MMRB which provides under the heading 'item coverage' a very comprehensive list, almost a narrative account, of the operations associated with, and covered by, each particular item. The object of such detail is certainty, but it has disadvantages similar to those of 'over-specification', for if anything is omitted from the item coverage a contractor will have strong grounds for asserting that an exclusion from so full a description must be deliberate and can only be re-introduced as a variation requiring additional payment. The effect is to limit the coverage afforded by implied terms and to transfer more risks from the contractor to the employer.

In order to keep the bill of quantities as clear and simple as possible, supporting detail may be presented in the form of supplementary schedules. Carriageway setting-out information, details of manhole sizes and pipe layouts, bar diameters and bending dimensions for steel reinforcement are typical examples of the contents of these schedules. They are usually separately bound and are amongst the most-used items in the site office. Sometimes the schedules are described as being 'for the assistance of tenderers' and are not formally incorporated into the contract documents, but this is a questionable practice which ignores the essential information they contain. The schedules should be listed amongst the contract documents, thus ensuring that the requirements of the contract are properly presented without the need for further explanation.

Interpretation and explanation

The contract documents can never describe the components of the works with absolute precision. Indeed, not only would such an attempt be doomed to almost certain failure, but it would also expose the employer to the risks of 'over-specification' whilst denying him the benefits of flexibility. By offering a degree of freedom of choice in method, plant, types and performance characteristics of materials, the employer allows the contractor to decide for himself how to make the best use of his own resources and where to seek supplies and facilities. These advantages are secured at the cost of some uncertainty, because the more flexible the expression of a requirement, the more open it becomes to different interpretations. The resident engineer and his staff have the duty of ensuring that work is carried out in accordance with the contract documents and will have to make frequent decisions on the meaning of various provisions or their relevance to particular circumstances. The inclusion of dimensional tolerances, the use of grading 'envelopes' for granular materials, the qualification of specific suppliers' or manufacturers' products with the phrase 'or similar approved' are just some examples of the ways in which the contract documents are prepared so as to leave room for flexibility. In such cases interpretation of the requirements is a necessary part of making the contract work.

The contract drawings for a bridge on a county council road contract included a schedule of the types and sizes of bearings required together with certain performance parameters. Two firms were named in a note accompanying the schedule which read: '. . . bearings to be supplied by X or Y Ltd or other similar approved manufacturers'. The contractor proposed a firm whose products satisfied the requirements in the schedule but were known to have a significantly shorter working life than those of the stated manufacturers. The resident engineer would not approve the proposal. On his interpretation of the provision, a third manufacturer would not be 'similar' just because he produced bearings within the same range of performance, they would also need to be comparable in overall quality.

Decisions of interpretation are not always self-contained and the nature of civil engineering work prevents easy isolation of one component or operation from another. The supervisory staff will often have to consider how their interpretation fits in with other requirements of the contract, an exercise demanding sound engineering judgement and experience because the contract documents are rarely cross-referenced to conveniently point out the links between the many facets of a construction project. The only effective way of approaching the problem of interpretation, and this is also true of the process of approval and acceptance, is to view as a whole both the works and the contract documents which describe it.

The conditions of contract expect the contract documents to fit together as a whole, for they are required to be 'mutually explanatory of one another' and it is a duty of the Engineer or his representative to explain and adjust any ambiguities or discrepancies. The Engineer does not have to await a reference from the contractor before taking action and should proceed on his own account once he becomes aware of a difficulty. In any event, the Engineer can only act on a genuine ambiguity or discrepancy; when the contract documents are clear they must be applied, although one or other (or even both) of the parties may be dissatisfied.

The requirement to be 'mutually explanatory' eliminates any particular order of precedence amongst the various parts of the contract documents, of the kind expressed in an old rule of thumb which placed the conditions of contract above the specification; and both above the contract drawings. However, there are certain guidelines which can be called upon as general aids to interpretation:

(1) Obvious clerical or typing errors are invalid.
(2) Words govern figures, so where there is a discrepancy in a dimension or quantity given in both words and numbers, the words are taken as correct.
(3) Specifically stated dimensions (whether in words or figures) are to be preferred over scaling off the drawings.
(4) Additions or modifications to a standard document (e.g. special clauses added to the conditions or specification) are to be given greater weight than any unaltered provisions with which they may conflict.
(5) Discrepancies which cannot be reconciled should be explained in the way least favourable to the employer on the grounds that he had control over the drafting and should not benefit from his own carelessness.

Finally, it must be stressed that ambiguities and discrepancies are to be 'explained and adjusted', not eliminated by changing the meaning of the contract documents.

The conditions of contract provide for extra payment to the contractor where disruption or delay result from the need to explain the documents, or where such explanation amounts to a variation. This does not mean that every discrepancy or ambiguity will produce money for the contractor, and even if he is genuinely misled, when, say, work is shown on a general arrangement but omitted from a detail plan, the test will be whether the mistake was a reasonable one to make in all the circumstances and not whether the mistake actually resulted in a loss.

In the contract for an underground car park the bill of quantities included a substantial length of power cable of various types to supply ventilation fans, lights, etc. The services diagrams in the contract documents did not show all the cables correctly although elsewhere in the drawings the arrangement of cable ducts was properly depicted. Whilst the ducts were being installed the Engineer detected the discrepancy, advised the contractor that the bill of quantities was correct and explained the correct cabling arrangement. This was not a variation as there was no change to the amount of cabling stated in the bill nor to the nature of the operation covered by the item description and thus the rate for the work remained unaltered. Similarly, the early adjustment of the discrepancy involved no delay as cabling had not even begun, and no disruption as the duct layout required no alteration.

On a flood prevention scheme the quantity of sheet piling shown in the bill of quantities was greater than that to be found on the drawings. The resident engineer adjusted the discrepancy by indicating an extension of the sheet piles beyond the limit shown on the drawings. The length of piling originally depicted could all be driven with a rig set up on the river bank, but the extended section had extremely limited working space and included some very uneven ground. The contractor had to drive this length of piles with a rig mounted on a pontoon. The adjustment of the discrepancy thus amounted to a variation as it involved a change in the nature of the work. Delay and disruption occurred because of difficulties in obtaining the equipment at short notice. The contractor recovered his extra costs from the employer.

Site staff know the complexities of a construction project and are therefore in a position to appreciate the difficulties of covering every aspect comprehensively and without error. When the inevitable ambiguities, inconsistencies and discrepancies emerge they should not come as a surprise but neither should their potential for disturbing the contractor's operations be underestimated. In putting matters right, the first priority must be to determine the correct solution in engineering terms with the need to minimise disruption to the works a close second. Any temptation to protect the designer (and thus the employer) from the consequences has to be resisted.

Chapter 5
Programme and method

Long before they arrive on site, the contractor's staff will have spent considerable time and effort drawing up a timetable for the project and deciding the methods of working to be used to carry it out. The Engineer and his site team are also interested in programming and method, but for different reasons.

The contract period, the time for completion of the works, is stated in the contract documents. This is the foundation on which the programme will be built, a structure which must not only stand up, but perform a number of vital functions as well. To the contractor's agent this timetable represents his production targets. It is also important for its role in the organisation of plant and labour and the control of subcontractors' operations and suppliers' deliveries. It governs the vital matter of cash flow, for although the estimators can adjust their pricing strategy to assist the financing of the contract (for instance, through the device of 'front-loading' the tender by enhancing the rates for early operations at the expense of those to be executed later, thus bringing cash in quickly and minimising borrowing) it remains true that the best results come from tightly programmed, speedily completed jobs. To the resident engineer the programme is less of an operational tool and more of a measuring instrument. He will study the contractor's timetable to determine whether the durations are practical and represent a reasonable allowance for executing the various operations in a safe and workmanlike manner. It will be used as the basis for comparing real progress against the estimated progress and against the passage of the contract period, and for assessing the effect of any delays, however caused.

The contractor chooses his methods of working both to satisfy the requirements of the contract and to make the most economic use of available resources and supplies. The contractor's agent has to make maximum use of his plant and labour and must avoid wherever possible any excessive 'peaks' or 'troughs', particularly in respect of expensive pieces of heavy equipment. He must try to arrange continuous and concentrated employment for any specialised plant or trades, and seek to get the fullest benefit out of reuseable items such as formwork panels, trench sheeting and scaffolding. The resident engineer will examine these proposals to ensure that they contain nothing that could prove detrimental to the permanent works, either by failing to meet the requirements of the specification or because their interaction with some other proposal would result in an unacceptable, even if unintended, by-product.

The contractor's programme

Clause 14 of the ICE Conditions of Contract includes a requirement for the production of an approved programme. The requirement is mandatory: the contractor must prepare and submit his proposal and the Engineer must consider and pronounce upon it. The format of the programme is usually left to the contractor's discretion, but the Engineer may withhold his approval if the type of submission is inappropriate to the nature of the project. A bar chart presentation would suit a contract where a relatively small number of large-scale, self-contained operations dominate the programme: a rural road project, for example, with major earthworks and carriageway construction elements but having only a minor structural content. Network analysis is more likely to be justified on contracts where many activities, both large and small, have to be linked together as steps in the completion of the whole project: the construction of a maintenance hanger, for instance, involving piled foundations on site, extensive steel fabrication off site and special welding operations as the steel sections are erected in sequence. A combination of the two planning techniques might be used on a contract where both sets of characteristics are present, such as a rural motorway project which involves a steel railway bridge, allowing key operations to be represented simply on the main bar chart programme and to be covered in detail on supplementary networks.

Whatever method is used, the 'clause 14 programme' should represent the preferred timetable on which the contractor has based his estimate. Unless the contract provides for staged completion, with sections of the works to be handed over on specified dates, the content of the programme is entirely a matter for the contractor and should be judged on its engineering merits alone.

There is, therefore, nothing to prevent the contractor from aiming to complete in a shorter time than the contract period: indeed, there are many advantages. Completion ahead of time can bring financial gains to both parties to the contract, although not every employer obtains a real cash benefit from early commissioning (most public works schemes fall into this category), and some may be embarassed by the acceleration of their projected expenditure. On contracts which are very susceptible to the weather or other variables (for example, projects involving construction in tidal waters), aiming to finish early is one way of building into the programme the 'float' needed to accommodate the probability of disruption. Again, the phasing of the contract period relative to the traditional 'construction season' of March to October may also influence the contractor to choose an early completion date.

A major highway project had a contract period of two years starting in June. The contractor considered using the full allowance of time, leaving a significant part of the carriageway works for completion in the last three months of the contract. However, this programme would have involved the expensive re-mobilisation of pavement and surfacing equipment sent off site at the end of the previous autumn, temporary works to protect and drain the exposed sub-grade through the winter, and a tightly scheduled final phase which would rely heavily for its success on the early arrival of favourable spring weather. The contractor chose the alternative of employing extra resources and programming for completion by the second Christmas, before the onset of the worst of the winter. This 18-month programme left the last six months of the contract period to absorb any delays due to bad weather, failure to meet the target outputs or any other cause.

A compressed programme might be considered ambitious, but should not be rejected for that reason. Optimism, however, can be cause for concern: optimism about the weather, for example, as when earthworks are shown continuing through the winter months; or optimism about the nature of the site, as when the contractor gambles that a large 'provisional' item for the treatment of unstable ground will not be used and makes no allowance for it in his programme. Most serious of all, and by no means rare, is the deliberate telescoping of activities to manufacture a programme with a false early finish, for use in support of an exaggerated claim for delay should a suitable opportunity present itself. Programmes of this kind are sometimes described, with unnecessary politeness, as 'unrealistic', but whatever the label there can be only one reaction when such a manoeuvre is suspected — rejection.

Any assessment would be incomplete without considering how the contractor intends carrying out his proposals. The ICE Conditions of Contract state that 'a general description of arrangements and methods' must be supplied, together with any 'further details and information' as may reasonably be requested. This rather loose requirement is open to wide interpretation: from the bald assurance by the contractor that 'sufficient resources will be made available to carry out the programme', to the demand from the Engineer's staff for a detailed resources statement showing itemised schedules of plant, labour and projected outputs. Neither of these extremes is likely to be productive, and if the process of submission and approval is to be of any value it must be based on a more practical view of the duties and responsibilities of the participants.

The examination of the programme, which must be presented within 21 days of tender acceptance, normally takes place on site in the first few weeks of the contract. The resident engineer, consulting with the agent as necessary, will make his assessment in two stages, beginning with an inspection of the contents of the programme and the arrangement and relationship of its various operations. The aim is to detect errors in the logic or the basic assumptions upon which the order and linking of the activities depends: in simple terms, it is a search for mistakes of the kind that put beams into place before their supports are shown as complete.

In the 'clause 14 programme' for a motorway contract the following errors were detected:

(1) insufficient allowance for curing time between placing concrete in the roof of an underpass and the opening of a haul route across the structure;
(2) the specified period of notice required for the diversion of high-voltage power lines not fully included;
(3) an existing major drainage outfall cut off by excavation work before completion of the new outfall;
(4) the installation of cable ducts in the verges by another contractor (notified in the contract documents) not included in the programme despite its effect on adjacent operations;
(5) hot rolled asphalt wearing course shown continuing into winter months when local records indicate average temperatures below the minimum stated in the specification for this operation;
(6) supplementary network presented for a viaduct, but without some key operations such as the application of waterproofing to the deck.

In some cases the assessment will not proceed beyond this stage, either because of the agent's reluctance to supply firm details of 'arrangements and methods' or due to the resident engineer's dissatisfaction with the extent of the information provided. As far as the amount and scope of supporting data is concerned, the standard of what is 'reasonable' cannot be precisely defined, although the resident engineer must be able to form a considered judgement on such essential matters as the duration of any activity or the practicability of running certain operations simultaneously. If, in his opinion, there is insufficient information for him to do this, the resident engineer may give a qualified approval, accepting the programme 'as a sequence of operations only', and referring to the lack of co-operation which has prevented a proper evaluation. Any subsequent dispute over programming will revolve around the agent's submission and the resident engineer's assessment: a comprehensive record of all the information presented and the views expressed is therefore vital.

In most cases adequate details are presented to support the programme, thus allowing the second stage of the assessment to proceed. The durations of activities are examined, individually or in combination, in terms of the potential output of the plant, labour and sources of supply allocated to them. The emphasis must be on what is realistic and due regard has to be given not only to contractual restrictions (for example, a ban on night and Sunday working) but also to practical considerations such as the weather and the effect of local conditions on the efficiency of plant.

A 'clause 14 programme' included an activity which depended upon a supply of imported stone. The duration was based upon a rate of 7,000 tonnes per week and in the supporting detail it could be seen that the supplier had agreed to deliver 1,000 tonnes per day via a nearby rail depot. However, the depot was restricted by the terms of its planning consent to weekday operations only and thus the maximum delivery rate could not exceed 5,000 tonnes per week. The duration of the activity had to be increased by 40 per cent.

In a thorough assessment of the programme and the supporting detail or explanations provided by the agent it is inevitable that the resident engineer will need to pose searching questions and make critical comments. An attitude of healthy, but open-minded, suspicion is entirely appropriate to this part of his duties. It is equally certain that these questions and comments will be received by the agent with varying degrees of indignation, sometimes genuine, sometimes not. On balance, most contractors see the 'clause 14 programme' as a necessary evil and few agents would want to encourage discussion of it. Nevertheless, on this and many other subjects, experienced site staff know that there is no point in mincing words. The occasional risk of real offence is infinitely preferable, from the employer's point of view, to the potential dangers of silence.

Methods of working

The resident engineer needs to know details of the agent's proposed methods of working not only to make sense of the programme, but also to assess their effects on the permanent works.

His assessment covers both the immediate effects on the construction stage and the long-term implications for the performance and durability of the works in service. In the first instance, the resident engineer's task is to check that the proposals will meet any specific requirements of the contract, will provide a reasonable standard of workmanship and will not damage or interfere with other parts of the works already constructed or still in progress. In the second, the purpose of the assessment is to ensure that there is nothing in the proposed method, whether taken by itself or in combination with other proposals or factors, which could give rise to unreasonable defects, breakdowns and deterioration in the works during its normal design life. In this exercise the resident engineer must also decide whether the problems he might discover are due to failure to satisfy implied terms, in which case the contractor can be asked to submit different proposals, or to loopholes in the contract, in which case the employer must be advised of the difficulty and invited to choose between accepting the consequences or varying the works.

A road contract involved the excavation and placing of large volumes of chalk which was known to be at a high moisture content. The contractor supplied details of his programme, which was shorter than the contract period, and his method of working, which indicated the use of motorscrapers. The resident engineer pointed out the likelihood, confirmed by published technical reports and other widely-available evidence, that the use of such plant would produce severe degradation in wet chalk, resulting in a soft putty-like material. In view of the immediate problem of compacting chalk properly in this condition, and the subsequent difficulties of preparing an adequate foundation for carriageway construction, the contractor was told that his proposed method was unacceptable. He was advised to re-submit with different excavation plant or with a longer programme incorporating 'rest periods' to allow each layer to stabilise before the next stage of construction.

The specification for structural concrete contained no restrictions on the source of aggregates. When the contractor submitted details of his working arrangements it became clear that the batching plant would be using marine aggregates. The chloride salt content of the contractor's source was not found to be excessive, but the resident engineer considered the level to be high enough to cause maintenance problems in the long term. As the contract documents included no general ban on aggregates of marine origin, and no specific limit on chloride salt content, the resident engineer advised the employer that the contract should be varied to prohibit marine aggregates and that the contractor be reimbursed for the extra cost of using an acceptable source.

Similarly, in a contract for the supply and erection of pre-cast concrete units at a sewage works, the specification did not restrict the source of aggregates. But in this case the contractor's proposed method of working included the use of steam curing at the casting yard. This technique is generally accepted to be bad practice when high chloride salt levels, such as those typically found in marine aggregates, are present as it produces a high probability of early deterioration in the concrete. The contractor's proposals were rejected pending either a change to non-marine aggregates or the elimination of steam curing.

In two of the three examples given above, the rejection of the contractor's unsatisfactory methods is combined with a description of some acceptable alternatives. This has been done for clarity and completeness, but in practice the resident engineer would not be so helpful. There are two reasons for such caution. Firstly, most contract clauses follow the example of the ICE Conditions of Contract and make the obligation to approve the contractor's programme and methods a passive one. Although any rejection must be reasoned and the resident engineer must say why he will not accept the proposal, there is no duty to respond positively — the resident engineer does not have to say what he would accept in its place. Secondly, although resident engineers may have the use of some sweeping clause (such as clause 13 in the ICE Conditions of Contract) which allows them to instruct and direct contractors almost unreservedly, few are so rash as to take advantage of it. The ICE version of this clause is typical in including a provision for reimbursing the contractor if the resident engineer's instructions or directions result in extra cost 'beyond that reasonably to have been foreseen by an experienced contractor at the time of tender'. This test clearly lacks precision and is fertile ground for disputes. In the example of the pre-casting problem above, suppose the resident engineer had directed the contractor to change his source of aggregate. Whilst accepting that the combination of steam curing and marine aggregate was detrimental to the permanent works, the contractor might, nevertheless, argue that a change in the curing method was the cheapest solution for him and that he had been put to extra cost by the positive instruction to follow the alternative course of action.

It is consequences such as these which have given rise to that long-standing warning to resident engineers, 'never instruct the contractor on method'. Lawyers and claim consultants give this advice constantly, and do good business when it is ignored, but it is a pity that the industry has become so disputes-conscious that resident engineers have to be exhorted to adopt a negative rather than a constructive attitude. Sometimes the value of the passive approach (or at least its freedom from risk) is rated so highly that resident engineers are advised not to ask for details of the contractor's methods and arrangements, thus avoiding the possibility of any comment, even a plain rejection, subsequently being judged 'beyond that reasonably to have been foreseen'. This is surely a counsel of despair. The resident engineer is on the site to exercise his professional judgement, otherwise he is no more than a watchman. Sooner or later he will see what the contractor's methods and arrangements are, and will form an opinion of their acceptability. How much better to face any problems at an early stage when measures can be taken to solve them and avoid any detrimental effects.

The resident engineer must be diligent as well as prudent. When his inquiries disclose a proposed method of working which is unacceptable, the problem should be faced constructively and systematically. The basic steps of the process are as follows:

(1) The first stage is a formal rejection which refers to the appropriate clauses in the conditions of contract stating reasons, but contains no instructions.
(2) The resident engineer can then discuss the matter informally with the agent, considering the alternatives which might be offered and giving guidance on what would be acceptable.

(3) If there is no successful conclusion, and the agent puts his unacceptable method into use, the resident engineer must notify him formally that he is working without approval, quoting the appropriate contractual authority (clause 13(2) in the case of the ICE Conditions of Contract) but issuing no instructions.
(4) In the absence of any new discussions, and with the agent persisting in the use of an unapproved method, the resident engineer is now bound to take some action to safeguard the employer's interests and must either stop operations or instruct the agent on the method to be used.

Any instruction which has to be issued must be clear and strictly limited to replacing the unacceptable method or arrangement. The resident engineer will need to record all the details of its execution and any consequential effects he can identify, and generally to act as if a dispute had been notified, for that will undoubtedly follow.

Temporary works

On any civil engineering project the contractor has to carry out work which, whilst essential to the construction process, is not part of the finished product and is removed before completion. There are three main types of temporary works:

(1) diversions to enable traffic (vehicles and pedestrians) or services (such as power supplies and sewers) to find their way around or through the site without interfering with the works and without serious interruption;
(2) formwork to mould and support concrete in the desired shape during placing and initial curing;
(3) falsework to support formwork or other components until the structure is capable of withstanding its working load.

Diversions are often included as items in the preliminaries section of the bill of quantities. Usually the contractor's freedom of choice is limited to some degree by indicating approximately the routes to be followed, by laying down minimum standards for construction or by specifying the general phasing of the temporary works with the main operations. In some cases the diversion arrangements are fixed during the design of the project, as a condition of planning consent, for instance, or in order to secure the approval of a landowner, and are set out in the contract documents in complete detail. Formwork, too, is normally included amongst the items in the bill of quantities. The specification may provide descriptions of the various classes of finish which have to be produced but unless some special effect is required the design of the forms, their means of connection and immediate support are all left to the contractor. Falsework design depends heavily on operational factors such as the speed with which it is to be erected and dismantled, the amount of space available and the degree of access required. Added to this may be the contractor's wish to get his falsework 'off the shelf', by choosing from amongst the many proprietary systems available one which particularly suits his needs or which he has used successfully before. These are powerful arguments for leaving falsework entirely in the contractor's hands, with no bill items or particular specification requirements. When a new construction technique is being introduced on the project, or the design is based on the use of a particular kind of falsework, the contract documents may provide full details and working drawings, but this is exceptional.

Concern over the dangers of falsework collapse led in 1975 to the Health and Safety Executive publishing the 'Bragg Report' (officially entitled the Final Report of the Advisory Committee on Falsework) and finally, in 1982, to the Code of Practice for Falsework (BS 5975). Most contractors have taken these recommendations to heart and follow them without the need for any prompting, but the main points — as far as site operations are concerned — can be summarised as follows:

(1) The contractor should give the duty of overseeing all activities relating to falsework on the site to one member of the staff, the temporary works co-ordinator (TWC).
(2) The TWC examines the brief issued to the falsework designers to ensure it is adequate and accords with actual site conditions.

(3) The TWC satisfies himself that the design is properly checked (this does not mean that he has to carry out the check, but he must see that it is done by a competent person) and then passes it on to the resident engineer.
(4) The resident engineer should examine the design thoroughly and give his consent, with or without modification (although this 'double check' does not take away the contractor's responsibility, the resident engineer should treat it seriously).
(5) The TWC manages the delivery and assembly of the falsework, checking the components and the method of erection against the design drawings.
(6) The resident engineer inspects the falsework as it proceeds, passing on any comments to the TWC for action.
(7) Loading of the falsework only takes place with the authority of the TWC, who must be satisfied that any faults detected during erection have been rectified.
(8) Subject to compliance with the requirements of the specification for strength tests or curing time, the falsework is struck under the direction of the TWC.

Amongst the advice regularly pressed on resident engineers is, 'leave temporary works to the contractor as they are entirely his responsibility'. Apart from the rare exceptions when designed temporary works are included in the documents, this advice is sound but misleadingly over-simplified.

In the past, conditions of contract were often ambiguous or contradictory, or both, about the requirements for temporary works. The old Fourth Edition of the ICE Conditions of Contract was no exception, containing loosely-drafted definitions and mixing references to the collective 'Works and Temporary Works' with those covering specifically 'the Permanent Works' and 'the Works'. This lack of clarity fostered a general view that temporary works only came within the resident engineer's jurisdiction if they were specified in the documents or were in danger of causing actual detriment to the permanent works.

The Fifth Edition clears away the uncertainty by bringing temporary works specifically into the definition of 'the Works' and so under the authority of the supervisory staff. The resident engineer can therefore seek a description of the 'arrangements and method' for any temporary works, may reject them as unacceptable and instruct different ones. Where his action results in additional cost or delay 'beyond that reasonably to have been foreseen' he must certify extra payment.

Methods of working leading directly to the completion of elements of the permanent works are easier to assess because a straightforward comparison can be made with the requirements of the specification or the drawings and any link between method and potential detriment is relatively simple to establish. Temporary works, by their very nature, separate cause from effect by at least one stage and so increase the problems of detecting and eliminating unacceptable features. Further difficulties can arise when a proposal is rejected not because of its detrimental effect on the permanent works but because, for example, it will restrict some essential access to the site, or cause danger to personnel, or damage adjacent property. Here the resident engineer has to proceed with caution, seeking first to arrive at an acceptable modification through negotiation, but remembering that recent developments in the law of negligence make prevarication or inaction a very risky course, from his own point of view as well as the employer's.

The falsework to provide temporary support for a post-tensioned concrete bridge was a proprietary system with some additional items supplied by the contractor. It was rejected on two grounds:

(1) The working platforms, to be made up from scaffold tubing and cantilevered out from the main framework, were inadequately supported and factors of safety calculated for the likely loading of men and equipment were dangerously low.
(2) There was no opening in the falsework to allow site traffic to pass beneath the structure thus, despite proper notice in the documents, preventing another contractor gaining access to part of the works.

The contractor re-submitted his proposal with stronger working platforms and an opening to accommodate site traffic. Because the opening was very narrow and construction vehicles might strike the adjacent formwork, the resident engineer instructed the following additional modifications:

(1) crash barriers to be erected on both sides of the track through the opening and extending clear for 10m on each approach;

(2) extra cross-bracing of scaffolding poles to be fitted in addition to those recommended by the falsework manufacturer.

Both were carried out but the contractor successfully claimed reimbursement for the second as an 'unforeseen' extra.

In general, the design of temporary works remains the contractor's responsibility and no approval from the resident engineer can alter this. There are, however, two major exceptions to this rule. The first occurs when the contractor is supplied with design criteria or calculations relating to the permanent works to assist him in formulating his proposals, in which case the employer is liable for any inaccuracies and their consequential effects. The second involves temporary works which are detailed in the contract documents and for which the employer is responsible, provided that proper skill has been used in their construction.

Chapter 6
Time and the contract

Time and the contract are inseparable. From the date of commencement, after which the contractor must make a start 'as soon as is reasonably possible', time begins to run, and its passage regulates every phase of the works. In almost every facet of supervision, the calendar or the clock plays a major role: the completion date, the various periods of notice set out in the conditions of contract and the specification, the curing time for concrete, delays, contemporary records, site diaries, output, delivery rates, — the catalogue seems endless. The site staff are involved in all of it, but four aspects are important enough to warrant special attention: the assessment of progress; the granting of extensions to the contract period; suspension of the works; the certification of completion and the arrangements for maintenance.

Progress

Once work on site has begun, the resident engineer and his staff will set up their procedures for monitoring the contractor's progress and any external circumstances which might affect it. This exercise has two purposes:

(1) comparing actual durations and resources with those in the programme to assess the accuracy and practicability of the contractor's assumptions;
(2) collecting information on the reasons for any delays, however caused, and their effects, both direct and indirect, on the progress of the works as a whole.

Regular progress reports provide basic data for assessing the contractor's entitlement to an extension of the contract period, and determining what its length might be. In evaluating the cost of extra work or other variations the resident engineer will look to these reports to demonstrate any effects on progress and to confirm the general level of output being achieved by the contractor on comparable operations. The investigation of claims for delay or disruption is one of the most complex tasks undertaken by the site staff and a continuous review of the progress achieved on, and the resources committed to, the various operations is essential to clarifying the issues involved. Of course, progress reports are not the only evidence the resident engineer will need, and in every instance they must be linked with supporting details from weather records, site diaries, delivery lists and other sources. Nevertheless, they represent the single best summary of the development of the contract, and act as a vital reference source both during construction and afterwards. They should be produced at regular intervals, and on time.

However much the resident engineer and the agent wish to avoid a formal dispute, there is no doubt that the record of progress is built up with this eventuality very much in mind. It is vital, therefore, that the resident engineer's information is collected and presented in a systematic fashion, certainly with full details of the relevant calculations or assumptions, and preferably with the agreement of the agent. This latter objective is not always easily achieved. The first obstacle is the natural optimism with which all agents are born, and which has been sustained by a succession of contracts, although never the one currently in progress, on which everything has gone to plan without the slightest interruption. Experienced resident engineers observe, without surprise, how this optimism becomes conspicuously extravagant when applied to the prospects of any operation subject to delay or disruption, provided the agent does not consider himself liable.

A more serious problem is how progress can be measured. It is sensible to make the review coincide with the monthly valuation, so, comparing the sum passed for payment with the total in the bill of quantities for that operation is an obvious solution. Although this can work well on self-contained items, most progress assessments are broadly based and relate to combinations or groups of items: the contractor's pricing strategy, for instance, or the presence of an item which includes some very expensive materials can bias the result and make it unrealistic. Another approach is to compare elapsed time with programmed time. This can also provide a reasonable approximation, but it is only accurate when applied to self-contained operations which are carried out as one uninterrupted activity. Combined or grouped operations may include periods of inaction or part-time working so that difficulties arise in reconciling the amount of productive time with the overall time allocated to that activity.

The drainage work on a new sports centre involved the replacement of a large diameter foul sewer and the reconstruction of a number of existing connections. The laying of the new sewer, being an early operation and requiring costly reinforced concrete pipes, carried a high rate to assist in the generation of cash flow; it represented over half the overall value of the drainage work and was programmed to take five weeks. The removal of the old sewer and the connection work were given a duration of ten weeks and carried relatively low rates. After completion of the new length of foul sewer and before any other drainage had begun the Agent proposed that the drainage operation be assessed at 60 per cent complete. The resident engineer's figure was 40 per cent.

A number of different operations made up the excavation element on a car park contract. The overall duration was eight weeks. After excavation had proceeded continuously for four weeks a progress review took place at which the agent stated that no excavating would be done for the next two weeks and thereafter half the previous resources would be returning to complete this work in a further two weeks. Calculating on the basis of working-weeks at a standard level of resources, the agent suggested that the excavation operation should be reported as 80 per cent complete. The resident engineer, basing his assessment simply on time elapsed, wanted the figure reduced to 50 per cent.

If the review shows that work is falling significantly behind schedule or that the contractor is not following the programme, then the resident engineer can demand a revision. Whilst the ICE Conditions of Contract are specific about the right to a revised programme, not simply a new network or bar-chart, but one 'showing the modifications to the original programme necessary to ensure completion', they are silent about supporting detail. This does not mean that the contractor need not produce any details of revised resources, and can merely submit the new and unsupported programme, for such an obligation can reasonably be considered an implied term of the contract. The resident engineer may justifiably argue that, in the absence of any details of how the original arrangements are to be changed, 'the modifications . . . necessary to ensure completion' have not been demonstrated, and so may reject the revised programme.

Whether seeking a new programme or drawing the agent's attention to slow progress, the resident engineer must take care to ensure his actions and comments are not misinterpreted as a formal instruction to accelerate the pace of work. Clause 46 of the ICE Conditions of Contract gives the resident engineer the authority to ask for proposals to expedite progress where there is a danger of the completion date not being met, but this is a power to be used with caution.

There are circumstances where the timing of operations can be fixed accurately enough to predict the earliest possible finish date with confidence, but most construction work is so complex and subject to external factors that estimating completion is a very uncertain business indeed. It is most unlikely that the agent will be prepared to admit that his plant and labour are working inefficiently and will therefore offer to improve output without changing resources, so any proposal to expedite progress is going to involve extended working hours or additional plant and labour or both, and consequently a claim for extra costs. The resident engineer must, therefore, be very sure of his ground before taking action. Even if he does not issue any instructions, he may confuse the situation to such an extent that 'time is set at large', the result of which is the elimination of the fixed contract period and with it the employer's right to damages for late completion.

It is also essential, if the contract period is to hold good, that the contractor is given possession of the site on time. If the contractor is prevented from occupying the site (because land purchase has not been completed) or cannot operate free from restrictions (because another contractor has not finished an earlier phase of the work) under clause 42 of the ICE Conditions of Contract he can claim for extra costs and will be entitled to an extension. If the delay is excessive or particularly disruptive, time may be 'set at large'.

Extensions of time

The basic timescale against which the engineer's staff assess the programme and evaluate progress is the contract period. The contractor can set out to reduce the time for the construction of the works but, except under special circumstances, he cannot extend this period.

Late completion normally involves the employer in some loss. This may be in the form of a direct loss of income when a new warehouse or factory is delayed, an indirect loss of benefit to the community when the opening of a sewage works or motorway is held up, or the cost of disruption when some subsequent work, the next phase of the project, for instance, has to be postponed. The employer protects himself against the potentially complicated and time-consuming task of determining the actual loss by stipulating a figure in advance and incorporating it into the contract. This sum is known as the liquidated damages. It is predetermined, accepted by both parties and expressed in cash terms. Although it is impossible to predict with any precision at the start of a contract what losses the employer might suffer at its end, which could be some years away, the liquidated damages must be calculated as realistically as possible and cannot be set up as a 'penalty clause', for this is illegal. Furthermore, once accepted as a term of the contract, these damages cannot be varied and whether the actual delay to completion means that the employer suffers a greater loss (because he underestimates the profitability of a new warehouse), or no loss at all (because a change in business plans prevents the opening of a new factory), the agreed sum is payable.

The usual arrangement is for liquidated damages to be fixed at so much per day, per week or per month. When the due date for completion arrives and the works are unfinished, the employer must be informed so that he, and not the Engineer, can deduct the appropriate figure from the contractor's payments. Once substantial completion is achieved, the employer is then notified so that deductions can be stopped.

All this depends upon the pre-condition that either the contractor has been notified of any extension of the contract period to which he may be entitled or, if no extension is justified, both the employer and the contractor have been informed that the specified completion date is to apply.

Clause 44 of the ICE Conditions of Contract provides a comprehensive three-stage procedure for assessing extensions, a responsibility considered so important that it is given solely to the Engineer and cannot be delegated. Although it would be unrealistic to expect the Engineer to act without consulting his supervisory staff (the resident engineer is often asked for a report and a recommendation) he must nevertheless apply his professional judgement and not merely 'rubber stamp' a decision made on site.

The first stage is the interim assessment which the contractor can request at any time during the progress of the works and is intended to produce a prompt, but not a final, decision. When the contractor has made an application and supplied full particulars, the Engineer must notify him formally whether or not an extension is to be granted and, if so, what its duration will be. The second stage is the assessment at the due date for completion. At this point the Engineer must consider all the circumstances and any representations from the contractor and either review his interim assessment, or, if no previous consideration has been given to an extension, make an assessment now, and decide whether the completion date should be amended. This is the essential pre-condition for enforcing liquidated damages and the employer and contractor must be informed of the outcome. The third stage, which coincides with the issue of the certificate of completion, is the final determination of extension. This is the Engineer's last opportunity to examine, or re-examine, the situation and fix the extension of the contract period. He cannot reduce any extension already granted.

There are two main grounds for granting an extension: 'contractual' and 'exceptional'. A significant variation to the contract or increase in quantities constitutes 'contractual' grounds, so does the activation of any of the contract clauses which provide for assessment of delay (late issue of contract documents, for instance, or an instruction under the 'satisfaction' clause). The use of a provisional sum is not a 'contractual' reason for an extension as long as the extent of the work actually required does not markedly exceed that stated in the bill of quantities (another good reason for making a genuine estimate of the quantity involved in a provisional sum).

'Exceptional' grounds are less easy to define. The so-called 'excepted risks' of war and other forms of violence, against which the contractor does not have to indemnify the employer for damage to the works, are prime candidates, but clause 44 makes a specific reference to a source of disruption which is one of the contractor's risks: weather. The form of words 'exceptional adverse weather' gives the Engineer a guide as to how he should exercise his judgement. His evaluation must concentrate on the severity of the cause rather than the effect. Therefore, although contractors sometimes consider that any bad weather entitles them to an extension if it halts or seriously disrupts progress, a severe delay occasioned by bad weather is not in itself grounds for an extension unless the weather was 'exceptionally adverse'. A week of deep snow in January can disrupt the works badly and may well be 'adverse', but in midwinter is it 'exceptional'? A day of rain in the summer is not unusual, but if it is heavy enough to flood the site and its approaches for the rest of the week it may be 'exceptionally adverse'.

The weather is not the only source of 'exceptional' grounds for an extension. Clause 44 refers, less specifically, to 'other special circumstances of any kind whatsoever'. In exercising this general discretion the resident engineer must bear in mind the standard set in clause 44 for the particular case of bad weather and see if some equivalent characteristics can be identified in the circumstances under consideration. A typical example might be an industrial dispute affecting the supply of diesel fuel, 'special circumstances' certainly, but not of an 'exceptionally adverse' nature if it only lasts for two or three days. A strike which disrupts deliveries over a period of months, however, might well qualify.

Having decided that an extension is justified, the engineer has to determine its length. The contractor must supply 'full and detailed particulars' but it is not the agent's subjective assessment of the time he has lost which governs the evaluation of any extension. The standard of performance must be the objective one set by that elusive (some unkind resident engineers might say mythical) creature, the 'reasonable, experienced contractor'. It is not the delay actually suffered by the contractor on site that determines the length of the extension, for he may be inefficient or slow in his response to the problem, but the Engineer's professional evaluation of the time which would have been lost despite the exercise of reasonable skill and care in minimising the disruption.

Wet weather, amounting to a total rainfall of 85mm, persisted for a 15-day period in August. The site, on a 'greenfield' industrial estate, became saturated and plant could not be moved for a further ten days. The contractor requested a 25-day extension. The resident engineer obtained local records for the previous ten years from the Meteorological Office and found that in an average August there had been five to six wet days and 20 mm of rainfall. Checking his own site records it became clear that the slowness of the site to drain at the end of the wet period was partly due to the build-up of water in a large ditch which ran through the site. The contractor had allowed this important drainage path to become blocked with spoil from his excavations. After considering this information the Engineer made his interim assessment and the contractor was given a 15-day extension for 'exceptional adverse' weather: ten days for the direct effects of the above average rainfall and five days for the consequential delays which would have resulted even had the ditch been kept clear.

It should not be imagined that, having three stages of assessment, the Engineer has three chances at getting the right answer and so need not be too careful about his first attempt. As it is not permitted to reduce the extension at any subsequent stage, an excessive interim assessment will stand for the rest of the contract and so deprive the employer of the proper level of liquidated damages. Potentially more serious, however, are the consequences of an under-estimate. Although there are two further opportunities to increase the interim assessment, these are late in the life of the contract and the contractor will have based his conduct of the works on the Engineer's initial response. If the assessment is significantly low there will be two serious results:

(1) the contractor has effectively been instructed to expedite the works and will have good grounds for claiming the cost of the additional resources employed in accelerating his progress;
(2) there is a strong possibility that 'time will be set at large' and the right to liquidated damages lost.

In fairness to both the contractor and the employer the interim assessment must be as accurate as possible. The subsequent stages should be treated not as opportunities to refine a guess into an estimate, but as compulsory reviews in which the relevance of any newly acquired or revised information can be examined. If there is no change in the evidence, there should be no need to alter the extension.

The view is sometimes expressed that an Engineer who has granted an extension for bad weather can, in the later stages of assessment, take into consideration the conditions during the whole contract period, the implication being that he should 'balance the books', with exceptional adverse weather representing a credit to the contractor and exceptional favourable weather appearing as a debit. The temptation is a strong one, especially if a period of good weather allows the contractor to improve on his programme, but there is no support for this view in the conditions of contract, indeed, if the good weather came after an interim assessment it could not be taken into account as this would involve reducing the extension. Furthermore, it is quite unjust to penalise the contractor for good fortune which benefits not only him but also the employer and the contract as a whole.

Granting an extension, therefore, is a serious business, but one which is often made into an unnecessary crisis by the mistaken belief that every extension produces a claim. Certainly, where the reason is an act of, or on behalf of, the employer (a substantial variation of the works, for example) or some fault within the employer's liability (such as failure to give full possession of the site) then the contractor may be entitled to extra payment. On the other hand an extension may be granted for a cause, albeit exceptional in its extent, which is one of the contractor's risks (such as bad weather or an industrial dispute). Under these circumstances the employer cannot deduct damages for late completion but neither can the contractor claim for the expense, however great, he may have incurred in overcoming the problems and finishing within the revised contract period. Indeed, where contractor's risks are involved the only way a claim can arise is if the Engineer delays unreasonably the granting of an extension or makes an unfairly low assessment.

In a piece of loose drafting which occasionally causes confusion on site, clause 44 of the ICE Conditions of Contract allows the Engineer, 'if he thinks fit', to grant an extension 'in the absence of any such claim'. It cannot be the intention of the ICE Conditions of Contract to encourage the Engineer to present the contractor with specific but unsought offers of an extension. To do so would be to undermine the employer's interest in prompt completion and to fail to act 'fairly as between the parties'. Furthermore, to give the Engineer this power would interfere with the general obligation placed on the contractor elsewhere in the ICE Conditions of Contract to report any claim as soon as he becomes aware of its existence. The anomaly can be explained by interpreting 'claim' not as a mere request for an extension, but as the request together with its supporting particulars. This would mean that the Engineer can 'if he thinks fit' consider and pronounce upon a request for an extension although it is presented without proper information. In practice, the Engineer should adopt a cautious approach to any request which the agent is unable or unwilling to support with proper details. The result may well be an underestimate but the contractor is in no position to complain if he has failed to supply the necessary details.

It is possible that an event might occur which is considered by the site staff to justify an extension, but the contractor takes no action whatsoever to claim it. In this unlikely but delicate situation the resident engineer should report back to the Engineer, whose response will take into account the nature of the event in question. If it is one of the contractor's risks then it should be assumed from his silence that due allowance has been made in his programme for such an eventuality and no further action is necessary. If the possible source of delay falls within the employer's liability, the contractor should be asked formally whether an extension is required. An affirmative answer would have to be accompanied by supporting details; a negative response closes the matter, although the contractor could reconsider and make a formal request later, in which case the employer would not be liable for the consequences of the delay in granting the extension.

Suspensions

Clause 40 of the ICE Conditions of Contract confers on the Engineer, or his delegate, the authority to suspend the contractor's operations anywhere on the site and even to bring the whole of the works to a halt. Few supervisors will see this power exercised, not because it is unnecessary, but because only a brave and confident resident engineer (or a very foolish one) would consider such action, and then only in exceptional circumstances.

If a suspension is to be ordered, there have to be very good reasons and the best are likely to be supplied by:

(1) the possibility of a safety hazard;
(2) the probability of serious damage to the works.

The careful distinction in the wording is deliberate. The law now imposes a strict duty of care in respect of dangers to employees and the general public (there is, of course, a moral obligation too) and a resident engineer need only be reasonably sure that a hazard is present or developing to consider suspension as a genuine option. Where his concern is for the integrity of the works, a resident engineer needs to be much more certain that real damage is going to occur or that serious defects will result, affecting the proper operation or durability of the finished product.

Two main causes of these situations are bad weather and the contractor's method of working. The former is the least contentious, in relative terms, as the contractor will at least be eligible for an extension of time. The latter means an inevitable dispute as it goes to the heart of such highly sensitive matters as the contractor's competence and good faith. It must be stressed that where bad weather is concerned it is not enough that progress is very slow or that the site is in a mess, it is not even enough that 'exceptional adverse' conditions exist which warrant an extension, a suspension can be justified only when to continue would be dangerous or seriously detrimental to the works. Similarly, the contractor's operations must represent a safety risk or a significant threat to the works, before the resident engineer should contemplate stopping them. Inefficiency, an unprofessional attitude, failure to comply with a provision of the specification, in themselves these are no grounds on which to suspend the works. After all, financial loss is the agent's business, the employer can resolve never to use that particular contractor again and bad workmanship or materials can be removed.

Due to inadequate setting-out and poor control of plant, a contractor building a large clay embankment for a motorway interchange began by substantially under-filling one side. Despite warnings from the resident engineer's staff the mistakes were not put right and work continued to full height. Compaction was satisfactory and the embankment was stable. The resident engineer instructed the agent to submit proposals for rectifying the error and refused to certify any payment for the fill. No suspension was ordered and later the contractor widened the embankment by benching-in the additional material, a costly and time-consuming operation carried out entirely at the contractor's expense.

A contractor had completed the first lift of a concrete water tower. The laboratory cubes indicated that the concrete was significantly below the specified strength. The agent suggested that the cubes were unrepresentative, offered to take cores from the placed concrete to ascertain the field strength, but indicated his intention to proceed with the second lift before this sampling and testing could be completed. The resident engineer agreed to consider the results from cored samples but rejected the proposal to continue with the next lift in anticipation of the outcome. His concern was not only for the damage the works would suffer from overstressing the first lift, but also for the danger to the men engaged on placing the concrete or other nearby operations if there should be a collapse. The agent was given formal notice that his proposed method of working was unacceptable and instructed to submit new proposals. No answer was received and early the next morning truck-mixers began to arrive from the local concrete plant. The resident engineer ordered the suspension of the concreting operation.

The consequences for the resident engineer of an error of judgement or any mishandling of the situation are awe-inspiring. An incorrect or invalid suspension not only attracts the cost of idle plant and labour and of subsequent acceleration, but also most agents will be able to demonstrate, whether on a sound basis of recorded fact or through an imaginative review of their programme, that almost every instance of delay and disruption later in the contract can be traced back to that original, mistaken suspension.

Nevertheless, despite the professionalism and commonsense of most contractors and their agents, there are occasions when discussion and persuasion are not enough and a suspension must be ordered. Once the order has been put into effect, the resident engineer will be keen to see the difficulties overcome and the stoppage brought to an end. This will involve close consultation with the agent to determine the necessary action and a continuous careful assessment of the need to maintain the suspension. Because of the exceptional circumstances, the site staff need to maintain their records with special care and ensure that all discussions with the agent and his staff are formally noted. It is particularly important to make clear the origin of any proposals which might involve extra cost and to differentiate between instructions and suggestions.

No-one on site would ever wish to see work stopped for more than three months, but if such a disastrous situation were to be reached the contractor can serve notice that he requires permission to proceed. If in 28 days permission has not been granted and the resident engineer cannot show that the reason for the continued suspension is some default of the contractor, then the consequences are grim indeed and can amount to effective abandonment of the contract.

The moral for resident engineers is that almost anything, short of failing to discharge their duties, is preferable to a suspension.

Completion and maintenance

The ICE Conditions of Contract (clause 48) require the contractor to notify the Engineer of completion, and within 21 days the Engineer must either certify that the works are 'substantially complete' or issue instructions specifying what outstanding work remains to be done. As with the granting of extensions of time, the issue of the certificate of completion cannot be delegated, but the Engineer is, nevertheless, bound to rely on the site staff for an assessment and it is usually through the resident engineer that the agent makes his first approaches.

There is nothing in the ICE Conditions of Contract to restrain the agent from bombarding the site staff with premature notices of completion. In addition, clause 48 allows requests to be made in respect of 'any substantial part' which has been finished and brought into use even though there is no provision for sectional completion in the contract documents. The pressure from the contractor's side is understandable, but a rushed decision can result in problems: a premature hand-over may leave too much work for completion in the subsequent maintenance period, when resources are limited and the pace of operations has slowed down. Partial completion risks transferring to the employer the liability for damage when construction operations may still affect that part of the works.

The use of the typically vague phrases 'substantially complete' and 'substantial part of the Works' leaves a wide area open for interpretation and dispute. It is clear that minor outstanding items, surface finishes for instance, or the removal of certain types of temporary works, cannot generally prevent the issue of the certificate, nor can small defects, for the contract covers their completion or correction. The best test is whether or not the incomplete work impairs the safe and effective use of the works, and the easiest way of deciding is to list what has to be done and the consequences to the employer, and perhaps the public if they are to have access, of accepting the works in that state. If the list is too long or its contents too important, it will be needed to accompany the refusal to issue the certificate of completion.

A large single-storey warehouse had been completed apart from the installation of the internal ceiling panels and the repair of some defective components in the roof. Both operations represented only a small part of the value of the contract but required the erection of access platforms and staging and the use of power tools overhead. The platforms had to be dismantled and re-erected several times as the whole of the roof area was involved. The risk to the safety of employees working in the warehouse and the disruption of storage arrangements were likely to be significant so the employer could not, therefore, make use of the structure. The Engineer issued the contractor with appropriate instructions to complete the work in the roof and ceiling and withheld the certificate of completion.

The issue of a certificate for a part of the works depends not only on its completion but also on its being 'occupied or used by the Employer'. There are many occasions where this is done for the employer's benefit, as when a subway or footbridge is brought into use to re-open an important pedestrian route severed by the site or when part of a factory is completed in advance to allow the installation of machinery to begin, and these are clear cases for issuing a 'partial' certificate. Sometimes, however, it is the contractor who gains the advantage, for instance, by opening a section of road early with the intention of using it as a haul route. In such a case the contractor's right to a certificate is much less clear-cut, particularly where the contract documents do not just ignore the option of sectional completion but imply its rejection (in the example just quoted, by including an item in the bill of quantities for a traffic diversion lasting the duration of the contract).

There can be no firm rule for determining what constitutes a 'substantial part of the Works'. Used alone, 'substantial' is misleading for it indicates that any large element might qualify for a certificate. Later, in clause 48, the qualification that it must be 'used or occupied' clearly implies that the part of the works in question must be to a large degree self-contained or independent of the remaining, incomplete sections. Indeed, this is a more relevant test than any assessment of how 'substantial' the part of the works might be. To return to a previous example, a subway or footbridge in a large urban scheme might account for only a small fraction of the works in both size and cost but nevertheless warrants its own certificate.

The certificate of completion indicates that the contractor has discharged his obligation to execute the works to the satisfaction of the Engineer. Therefore, the site staff must take care to ensure that any minor work not started is excluded from the certificate and any incomplete work is listed. In addition, any obvious faults, what the law knows as 'patent defects', should be notified to prevent any future disputes over their cause, as the contractor is not liable for the repair of accidental damage which occurs after completion.

The contractor has another obligation: to execute the works in accordance with the contract. This covers the rectification of any hidden faults which are only revealed by the passage of time, the 'latent defects' which have escaped the attention of the resident engineer and his staff. The period of maintenance, which is usually of twelve month's duration and starts to run as soon as completion is certified, is intended to provide the opportunity for these faults, which exclude 'fair wear and tear' and any defects arising out of inadequate design, to be detected and put right.

During this stage of the contract the resident engineer supervises the completion of the outstanding items specified in the certificate of completion and carries out a comprehensive check to detect any faults which may develop in use or were overlooked earlier. A final inspection is undertaken in the last weeks of the period of maintenance and the findings submitted formally to the contractor, at the very latest within fourteen days of the expiry of the period. The items on the list must be completed 'as soon as practicable' and represent, as far as the Engineer can ascertain, the limits of the contractor's obligation in respect of the execution of the works.

Once this last list has been cleared, the Engineer can issue to the employer, with a copy to the contractor, the maintenance certificate, stating (in the form of words included in clause 61 of the ICE Conditions of Contract) that the contractor has 'completed his obligations to construct complete and maintain the Works to the Engineer's satisfaction'. The wording is suitably final and this marks the discharge of the agreement between the employer and contractor, but it is not an end to the liabilities between the parties nor the powers of the

Engineer. Time is still running under the contract in three ways:

(1) The final account must be submitted within three months of the maintenance certificate and, provided it is properly supplied with supporting data, must be followed within the next three months by the issue of the Engineer's final certificate. Any outstanding sums must then be settled within 28 days.
(2) For a breach of contract, which may involve some failure to carry out the works satisfactorily or some disputed payment, the parties have six years from the date of the breach (12 if the contract is under seal) to bring an action.
(3) For a breach concealed by fraud the limitation periods remain the same but time starts to run only from the date when the fraud is discovered (or from when a reasonably diligent person ought to have discovered it).

The limitation period for defects arising out of negligence (in construction, design or supervision) was established by the House of Lords' decision in Anns *v* London Borough of Merton (1978) A.C. 728 as six years from the date when the defect became discoverable. This offered the potential for 'open-ended' liability and in Pirelli Ltd *v* Oscar Faber and Partners (1983) W.L.R.6 the Lords redefined the start of the limitation period as the date on which the defect manifested itself as damage or some other detrimental effect, whether discovered or not. In view of the difficulty of determining such a time accurately, it is unlikely that the law will remain in this state for long. A fixed time limit for latent defects of perhaps 15 or 20 years from the date of construction may be the result, but in any event site staff can clearly become involved in disputes stretching back beyond the reasonable limits of their memories, yet another incentive for keeping the fullest possible records.

Chapter 7
Safety

The safety record of the construction industry is not good, two or three times a week a fatal accident is reported somewhere in the United Kingdom. In part this is because of the nature of the operations which require men to go into deep excavations or on to high scaffolding, to work adjacent to large and fast-moving plant or in noisy conditions where warning signals can go unheard, to handle heavy equipment or materials, to operate powerful tools and to use dangerous chemicals and compressed gases. The distribution of loading in a structure during the construction stage is often significantly different from that carried in its normal working life and so factors of safety can be temporarily reduced, with the potentially tragic results demonstrated by the box-girder collapses of the early 1970s.

Historically, too, civil engineering has existed in an environment where safety was not a prime concern. Born into the violent world of the Industrial Revolution, growing up in the most expansionist phase of Victorian capitalism and nurtured by the gospels of self-help and laissez-faire, the construction industry still shelters in some quarters that traditional and fatalistic view of safety as a nuisance which gets in the way of 'proper work'. The roots of this attitude can be found amongst many of the great engineering projects of the past, although none surpasses the consistent disregard for safety shown by the builders of the Woodhead Tunnel. Taking the Sheffield to Manchester railway under the Pennines, it was three miles long, was started in 1838 and took seven years to build. Thirty-two men were killed and 140 seriously injured during its construction, a proportionally heavier casualty rate than that suffered by the British Army at Waterloo. Although by no means the only source of injury, the crude methods of firing the blasting charges at the rock face accounted for many of these accidents. When an official inquiry suggested that the patent fuse was the only really safe technique, the resident engineer, Mr. Wellington Purdon, felt able to reply without apparent remorse, 'perhaps it is ... but I would not recommend the loss of time for the sake of the extra lives it would save'.

No-one would take such an attitude today, yet that legacy of irresponsibility and complacency is not without its effects. There are still agents whose enthusiasm for safety checks is limited to their use as a stick for beating awkward subcontractors, resident engineers who turn a blind eye to dangerous practices so long as the public are not at risk, organisations which print volumes of safety procedures but do not provide the equipment they describe and site personnel who genuinely believe 'it will never happen to me'. For most supervisory staff what is needed is not specific expertise in this wide and complex subject (the safety officer at headquarters can supply the technical support) but a firm grasp of the basic requirements of the contract and the legislation, and a reasonable amount of common sense.

Safety under the ICE Conditions of Contract

Clause 8 of the ICE Conditions of Contract makes the contractor's obligations perfectly clear. He must 'take full responsibility for the adequacy, stability and safety of all site operations and methods of construction', and the contract excludes from this obligation only matters arising directly out of the design or the technical specification as set out in the contract documents. As far as the execution of the works (and this includes any temporary works) is concerned, the resident engineer has no contractual duty to check whether any proposed meods of working are safe, nor does his approval imply that he accepts them as safe. Thus, the resident engineer's approval of the contractor's excavation arrangements covers only their general suitability for executing the permanent works. It is not an endorsement of the proposal as safe in principle, and certainly cannot represent acceptance in advance of the working practices, materials and detailed arrangements involved.

The absence of a specific duty to check for safety can mislead some agents into thinking that the resident engineer has no interest in this aspect and no right to intervene. This is a grave misinterpretation of the resident engineer's role and the requirements of the contract. Site staff take a keen interest in safety not only because, for reasons which will be covered later, they are under certain legal obligations to do so, but also because the products of unsafe work are rarely adequate or stable and are often the source of trouble and failure later.

Furthermore, a collapse or any other major accident can have a serious effect on progress, and most employers prefer completion on time to liquidated damages. The site staff are also aware that adjacent landowners or occupiers, and perhaps members of the public, may suffer damage or injury as a result of the contractor's failure to work safely. Last, but not to be ignored, is the harm that can be done to the employer's good name (and the engineer's professional reputation) as a result of his involvement in a project with a bad safety record.

As far as the right to intervene is concerned, the resident engineer is empowered to apply his test of 'satisfaction' to the contractor's work and has the corresponding authority to instruct any changes which he may consider to be necessary. As always, there may be extra cost for the employer if the instruction involves work 'not reasonably to have been foreseen'. In this context, however, it is not 'reasonable' for the contractor to assume that the cheapest solution is necessarily the right response. In assessing the level of precautions to be taken it is justifiable to consider the financial element, but only when balanced against the degree and probability of risk. A minor hazard, for example, would not necessitate the introduction of costly and elaborate safety measures, but the danger of a serious accident, even if relatively unlikely, would require comprehensive protection and a major risk coupled with a high probability justifies the most extensive and expensive safeguards — 'belt and braces' treatment.

The contractor's responsibilities under the contract are not confined to the safety of his methods of working. Clause 19 of the ICE Conditions of Contract requires him:

(1) '... to have full regard for the safety of all persons entitled to be upon the Site ...';
(2) '... to keep the Site ... in an orderly state appropriate to the avoidance of danger ...';
(3) 'to provide and maintain ... all lights, guards, fencing, warning signs and watching ... for the safety and convenience of the public or others'.

This means that the works, and especially its perimeter, must be kept secure and any dangerous features such as excavations, partly completed structures and stockpiles of materials, must be made inaccessible. When diversions run through the site, or access tracks are left open for delivery vehicles, special care is needed to ensure that the limits of safe areas are well marked so that no-one can stray into parts of the works where their unexpected presence will be a hazard.

A temporary plant crossing, controlled by traffic lights, had been set up to allow earthmoving equipment to cross a busy road. The usual signs found at temporary traffic lights were on display, but there were no other warnings or 'keep out' notices and no one in attendance. A van driver, visiting the site, without authority, to deliver a message to a friend, turned off the road at the crossing and found himself in the middle of the haul route, surrounded by clouds of dust from fast-moving motorscrapers. Seeking the way out, he stayed close behind a scraper and followed it back to the plant crossing. The lights on the haul route were at red, but there was no traffic on the road. The scraper driver concluded that the radar controller had failed to detect his presence and, knowing that the next machine was some minutes behind, decided to move slightly to activate the lights. He put the scraper, weighing 57 tonnes fully laden, into reverse and moved back 5 m. The van was crushed.

Works in urban areas particularly highway schemes, inevitably mean the introduction of potential hazards such as open excavations, projections and overhangs into areas used by the public. In situations of this kind the liberal application of lamps, flutter barrier, cones and warning signs, although very necessary, is not enough: there must be a physical barrier. For obvious reasons of convenience and economy, agents prefer to mark out a danger zone with easily moved, cheaply replaced items and avoid wherever possible the use of fixed, expensive fencing or guardrails, yet the latter are the only certain way of fulfilling the safety requirements. When the agent, faced with the resident engineer's demand for yet another length of temporary fence, points to the cones, flashing beacons and splashes of red paint with an exasperated 'you'd have to be blind to miss that lot', he is expressing in layman's terms the very test which a lawyer would use to decide whether he was liable in the event of an accident. Twenty years ago it was decided in Haley *v* London Electricity Board (1965) A.C. 778 that anyone making an excavation in the highway ought reasonably to foresee that unaccompanied blind persons might come along, relying only on the use of a stick to detect obstructions or hazards in their path. The fact that sufficient precautions had been taken to allow a sighted person to avoid the danger did not save the Electricity Board from liability and the judgement has set the standard of care at this very high level. The Disabled Persons Act 1981 has reinforced the need to provide for the safety of all members of the public and not just the fit and alert majority.

'Haley's case' forms part of the law of negligence, and is not mentioned in the ICE Conditions of Contract. Nevertheless it is a sure guide to the degree of care which the employer can expect from an experienced contractor and which the site staff can reasonably insist upon. Furthermore, it is important to note that the law imposes duties on the contractor beyond the requirements of the ICE Conditions of Contract, which appear to limit the contractor's responsibility to 'persons entitled to be on the site'. A strong line of judgements have established the principle that where it is reasonable to expect people to go, even as trespassers, then the person in control of that place must take sensible precautions to minimise the risk of injury. Another decision, in British Railways Board *v* Herrington (1972) A.C. 877, clarified the legal position by holding the Board liable for the injuries to a child who trespassed on to an electrified railway by climbing through a dilapidated fence. Children played in a meadow adjacent to the line and it was reasonable to expect that some would try and get on to the track. The Board had not taken suitable steps to prevent this happening. This duty of care where trespass is concerned is not limited to railway operators, it extends to the occupiers of other potentially dangerous places. When the principle is applied to a construction site it means that the contractor must either close it off entirely with a solid barrier (the way most urban building sites are treated), or fence the works at all points of access and mount a proper watch throughout the day (the usual arrangement on large sites in rural areas).

Safety precautions instructed by the resident engineer will always be subject to the usual provisions for payment if they are 'unreasonable', but there can be no good excuse for short-cuts or false economies. The employer, and the public, have a right to expect the agent to take an active interest in matters of safety and security, based on prevention rather than cure. Similarly, the resident engineer and his staff will be expected to give this aspect of site supervision just as much priority as the work of inspecting and checking the construction operations.

A new road was being constructed through woodland alongside a suburban housing estate. The site was enclosed by a 1.2 m high temporary fence of chestnut palings. There was no watchman at night or at weekends. Children were known to be climbing over and through the fence to play amongst the equipment and materials, and the resident engineer asked for regular patrols to be made outside normal working hours. Instead, the agent erected several warning signs and arranged for the fence to be checked before work finished each evening. The resident engineer reluctantly accepted these measures, but when drain construction began he became concerned and requested the provision of either a watchman or separate fencing to any excavation over 1 m deep. The agent gave an assurance that no excavations deeper than 1 m would be left open overnight, and again argued that a watchman was unnecessary. Although uneasy, the resident engineer did not press the matter further nor did he initiate any patrols of his own. During the weekend a small boy fell into an open manhole excavation, over 3 m deep and partly flooded. Although trapped for some time, his cries were heard and he was rescued by the police. Amongst considerable publicity, a watchman *and* temporary fencing to all excavations were provided at once.

Safety legislation

The first serious attempt to legislate for safety was the Explosives Act 1875. Thereafter, for over 80 years, a whole range of Acts of Parliament and Regulations have come into being in a remarkably piecemeal fashion. The construction industry became subject to such detailed and diverse provisions as the Electricity Regulations 1908 (and as amended 1944) and the Boiler Explosions Act 1922.

The culmination of this process was the Factories Act 1961 which attempted to lay down a general framework of statutory requirements for specific types of workplace, including construction sites. The Act authorised the issue of supplementary provisions to cover particular operations or hazards and those relating to civil engineering are known as the Construction Regulations. However, this legislation was not wholly successful and shared with its predecessors (many of which still remain in force) two major defects:

(1) the use of regulations to exercise control in detail tended to produce a complicated, poorly understood and often out-of-date system which remained as fragmented as before;
(2) the self-employed were not covered.

The Robens Committee of 1972 criticised the negative approach of external regulation and proposed a change of emphasis towards a positive, self-regulatory system. The result was the Health and Safety at Work Act 1974 which is aimed primarily at people and their activities rather than premises and processes. The earlier legislation is still in existence but the Act supplements it by imposing a new principle: all persons at work have a responsibility for safety.

Specifically, employers must provide, as far as is reasonably practicable:

(1) safe plant and systems of work;
(2) safe handling, storage and transport of goods;
(3) information and training;
(4) safe places and access to work;
(5) a safe and healthy environment.

Moreover, they must conduct their activities in such a way as to protect persons not in their employment who may be affected, which includes self-employed persons and members of the public. Employees, for their part, are required to take reasonable care of their own safety and that of others, and to co-operate in ensuring that all statutory duties are properly discharged. The self-employed, who are found in large numbers in the construction industry, are now obliged to carry on their work in such a manner that other persons are not exposed to health or safety risks and, in another new statutory requirement, manufacturers and suppliers must ensure that their equipment and products are safe in use. The Act is administered, and its provisions enforced, by the Health and Safety Executive.

The broad definitions of responsibility and the emphasis on self regulation mean that all the participants in a civil engineering project carry important and interlocking duties. The contractor, his sub-contractors, the site personnel, self-employed labour, off-site suppliers and fabricators all have obligations to their staff, to themselves and to each other. The Engineer and his staff are involved, not only because they form a working group but also because they are able to direct and instruct activity on the site. Section 4 of the Act includes a widely-phrased provision to cover persons who exercise any control over a place of work:

'(2) It shall be the duty of each person who has, to any extent, control of premises to which this section applies or of the means of access to or egress therefrom or of any plant or substance in such premises to take such measures as is reasonable for a person in his position to take to ensure, so far as is reasonably practicable, that the premises, all means of access thereto or egress therefrom available for use by persons using the premises, and any plant or substance in the premises or, as the case may be, available for use there, is or are safe and without risks to health.'

As well as the Engineer's supervisory staff, the employer and his personnel may also be said to have some degree of 'control'. Indeed, under clause 19 of the ICE Conditions of Contract, the employer is specifically responsible, as far as his agreement with the contractor is concerned, for the safety of the operations of any of his own workmen who may be on the site, and any other persons (another contractor, for example) who may have been brought in to share the site.

The Health and Safety at Work Act allows for the issue of detailed regulations (which have statutory force) and codes of practice (which do not), but at the heart of the new approach to safety is the requirement in section 2 for all employing organisations to draw up a written safety policy. This document, which must be kept up-to-date, is usually in the form of a short general statement supplemented by a series of specific procedures which deal with the various activities of the organisation or of particular functional groups within it. A contractor's safety policy would include procedures for the operation of plant, the erection of scaffolding, work in excavations and every other operation likely to be found on a construction site. The Engineer's organisation (which might be a government department, a public authority or a firm of consulting engineers) would issue a policy with procedures covering, for instance, the use of testing equipment, inspections in dangerous places such as on scaffolding or in trenches, surveying on the public highway and the supervision of visitors to the site.

Although the adequacy or otherwise of the Contractor's safety policy is a matter for the Health and Safety Executive, the resident engineer can reasonably ask the agent for a copy to satisfy himself that it is being properly implemented without risk to his own staff or the public. As for the policy and procedures of his own organisation, the resident engineer must ensure that they are read, understood and put into practice by the site team. By showing a responsible attitude to their personal safety obligations, the resident engineer's staff strengthen his position considerably in any disputes with the agent over the safe operation of the site.

Any failure by the contractor to carry out the requirements of the Health and Safety legislation is an illegal act, which could result in individuals being removed from the site (under clause 16 of the ICE Conditions of Contract) or amount to a breach of contract. Furthermore, the intervention of one of the Health and Safety Executive's inspectors can bring very serious consequences for the progress of the works and the people in control of it. The formal machinery of enforcement operates through three types of notice which are described in Figure 4.

To some extent the inspector's choice of notice will depend upon his assessment of the attitude of the person on whom it is served, as well as the hazard involved. In the case of an isolated mistake or fault, where he finds a co-operative and positive response, the inspector is unlikely to act severely. A persistent disregard for elementary standards of safety, however, will certainly result in a compulsory stoppage of work. In many cases the inspector relies on a verbal warning or a cautionary letter, and the notice procedure is kept in reserve. When the inspector does use his powers of enforcement, he is most likely to do so through an 'Improvement Notice' which sets a time period during which the offender must put matters right or lodge an appeal to an industrial tribunal. If there is a real hazard to people the positive step of prohibition may be necessary. The 'Prohibition Notice' is a very flexible instrument of enforcement: it requires no specific contravention of any Act or Regulation, the risk of personal injury is enough; it may be served on persons under whose 'control' a dangerous activity is being carried on, and this could conceivably be a resident engineer although the agent would normally be the addressee.

Whenever a notice is issued, it requires some action to be taken and once that has been done the notice is extinguished. The courts are not involved and there is no question of a 'conviction'. Notices are devices to stop hazardous activities on the spot and are not retributive measures. Nevertheless, this does not mean that criminal prosecution is ruled out. Whether or not a notice is issued, and whether or not it is properly complied with, the inspector may still choose to institute proceedings if the matter is sufficiently serious. The Act makes individuals liable, as well as the corporate bodies to which they belong, and thus the agent, the resident engineer and

Type of notice	Circumstances of issue: Legal contravention	Circumstances of issue: Risk	Effective date	Addressee	Typical example
Improvement	There must be a contravention of a legal provision (or have been one), which is likely to be continued or repeated.	No risk necessary	Specified in the notice (but not earlier than 21 days after issue)	Person contravening the provision	The idler rollers on a loading conveyor not properly guarded
Prohibition (immediate)	Not necessary	Must be an imminent risk of serious personal injury	Immediate	Person carrying on the activity or person under whose control the activity is being carried on	Tower crane with a damaged jib, likely to fail under load
Prohibition (deferred)	Not necessary	Must be a potential risk of serious personal injury	At any time specified in the notice	As above	Scaffolding under erection but not yet in use, with damaged and corroded components

Figure 4 The Health and Safety at Work Act notices

members of their staffs can be called personally to account. The tragic deaths from asphyxiation of four contractor's men in an inspection chamber at the Carsington reservoir site in 1983 focused attention on this liability. Insufficient safety and rescue apparatus had been provided and both the contractor *and* the three partners of the consulting engineer supervising the work were charged. All the defendants pleaded guilty and were given substantial fines.

Late one evening, a member of the resident engineer's staff was leaving the site of a new factory when he noticed that some scaffolding had been built up dangerously close to a set of overhead high-voltage power cables. The resident engineer's assistant estimated that the gap was small enough to put anyone who climbed on to the scaffolding at risk of electrocution. He suspected that children played on the site during the night and knew that some of the contractor's men would be making an early start next morning, working near the scaffolding. He searched the site but found that everyone had left; he tried to telephone the agent at his home but got no answer; he contemplated calling out the Electricity Board's local emergency gang to have the power shut down but dismissed this as an over-reaction.

Concerned, he resolved to get to the site early the following day to pursue the matter. When he arrived he found the gates closed, an 'Immediate Prohibition Notice' in force and one of the Health and Safety Executive's inspectors present. The inspector had passed the site by chance late the night before, seen the danger and called out the Electricity Board.

The contractor's agent, and the resident engineer's assistant, were summonsed.

Emergencies

The need to take urgent action to eliminate a safety hazard or prevent damage to the works may arise at any time during a construction project. The laws of probability notwithstanding, resident engineers will testify to the perverse frequency with which such crises occur at weekends, over holiday periods and at other times when the contractor's labour force is dispersed off site. It is essential that the contractor is in a position to discharge his obligations regarding safety and security, and so a small gang, including an experienced foreman and at least two machine operators, should always be on call to provide emergency cover. Both the resident engineer and the agent should arrange a rota of their key personnel to be on stand-by whenever the site is left unattended, with their telephone numbers circulated and also issued to the local police.

Clause 45 of the ICE Conditions of Contract waives the normal restrictions on night and Sunday working when such action is 'absolutely necessary for the saving of life or property or the safety of the works'. Clause 62 covers the situation where the contractor is unable or unwilling to carry out urgent repairs which 'by reason of any accident or failure or other event ... shall in the opinion of the Engineer be urgently necessary' and allows other workmen to be brought on to the site to do the job. This course of action can lead to serious disputes and should not be undertaken lightly. No criteria are given to help determine what reasonably amounts to 'urgently necessary', although it is suggested that the extract from clause 45 quoted above offers some guidance. There can be no doubt, however, that where a risk of serious personal injury is involved there must be no hesitation in taking whatever precautions are necessary to protect workmen or the public.

The catalogue of potential disaster is long and runs from unlit obstructions on the public highway through fractured gas mains to blocked railway lines. In every case there is one absolute priority which must govern the immediate response: the safety of people must come first.

Staff safety and welfare

The resident engineer's concern for the safety and welfare of his staff should not rest on his legal obligations alone, but on his responsibility as a professional entrusted with the management of people. He should ensure that the need for all safety precautions is properly explained to new or inexperienced personnel, that full details of all safety procedures are known to the staff and the equipment necessary to comply with them is available. He must set a good example and also see to it that visitors to the site, no matter how important or senior in the Engineer's organisation, also obey the safety rules.

There can never be a good reason for site staff putting themselves at risk to carry out their duties. If a trench is inadequately supported, if steelwork is without proper access platforms, if a vertical shutter appears unstable then no-one should go and inspect or supervise that work. Very occasionally, the contractor's men may fail to appreciate the need to make such dangerous areas safe for themselves as well as the resident engineer's staff. Accusations of obstructive behaviour or unreasonable caution will result, perhaps reinforced by provocative remarks about resident engineers who are 'nervous' or 'soft'. None of this can make any difference. The contractor is bound to provide 'every facility ... and every assistance' to the resident engineer and his staff in performing their duties, and this includes the provision of safe conditions. In their absence there will be no inspection, no supervision, no approval and no payment.

The personal safety and welfare of the members of the site team is also regulated by two often-ignored pieces of legislation. The site offices are covered by the Offices, Shops and Railway Premises Act 1963 which sets standards for working space, ventilation, heating and toilet facilities. Most importantly, the Act requires fire extinguishers and escape routes to be provided and regularly maintained. This duty must be taken seriously, for the typical site office is built of wood, roofed with bituminous felt and stands alongside a laboratory of similar construction, well stocked with inflammable liquids. More common than fire are the hazards which result in the steady stream of accidents, fortunately usually of a minor nature, which mark the progress of a contract. The Health and Safety (First Aid) Regulations 1981 apply to the site organisation in two ways:

(1) all offices with up to 150 employees must have a properly stocked 'First Aid' box and an 'appointed person' who, although not a trained 'First Aider', has a duty to 'take charge of the situation if a serious injury or major illness occurs in the absence of a First Aider'.
(2) all personnel working away from the main office in isolated and/or potentially dangerous situations must carry travelling 'First Aid' kits.

When accidents resulting in injury to any of his staff do

occur, the resident engineer must see that a full report is prepared, not only to stand as evidence in any claim of negligence which might follow, but also to satisfy the requirements of the Health and Safety (Notification of Accidents and Dangerous Occurrences) Regulations 1980. The Regulations place a duty on 'the reponsible person' who is 'in control of the premises' to report immediately to the Health and Safety Executive any accident causing death or major injury to persons in his charge or any members of the public, ('major injury' means, in broad terms, something requiring admission to hospital). The details should be entered in the site record book (normally the Health and Safety Executive's Form 2509), and within seven days a full written report (on Form 2508) must go to the Executive. Injuries which are not 'major', but result in more than three days' absence from normal work, need not be reported but must be entered in the record book. Notifiable 'dangerous occurrences', which are defined in the regulations and in terms of construction sites involve most collapses (including scaffolding and falsework) and accidents to lifting or excavating plant, must also be fully documented, but these are generally under the control of the contractor and so the agent is 'the responsible person'.

However, the welfare of the staff involves more than just completing forms and is not secured simply by providing the statutory accommodation and facilities. Construction sites are generally dirty, noisy, exposed to the elements and subject to long working hours. The resident engineer will want to obtain the best quality equipment, protective clothing and office fittings to ensure that the well-being, morale and efficiency of his team is kept at a high level and, for the same reason, he will recognise the importance of keeping the building properly cleaned and maintained. There may be jibes from the contractor about the contrast between their spartan conditions and the luxury at the other end of the site compound, but these seem to fall more easily from the lips of the contracts manager, down for the day from the lavishly furnished headquarters, than from the agent who knows the benefit of decent surroundings.

Safety, and welfare too, are largely matters of attitude,

habit and common sense. All the legislation in the statute book will be of no value if these aspects of site work are not treated seriously and as part of a process of education. Too often the same lessons are re-learned from first principles, at unnecessary cost and with little evidence of that practical instinct which an engineering training is supposed to foster.

One particular safety rule is discovered, and subsequently forgotten, on almost every contract equipped with site transport. It is repeated here with the small hope of saving one or two of the injuries which will undoubtedly occur as the process of re-discovery goes on: given no drop in level from one side of an open trench to the other, it cannot be crossed by a Land Rover, no matter how fast it is driven.

Chapter 8
On site: organisation and communication

In the language of psychology, the resident engineer's site team is a task-orientated, self-motivating small group. For once, the jargon is almost self-explanatory. Its task of supervising the contractor's operations is the sole reason for the existence of the team. It has no other competing functions or goals and its structure and membership reflect this singleness of purpose. Furthermore, the team must, to a large extent, provide its own driving force. Although there are many occasions when the contractor generates activity, by seeking approval to place concrete, for instance, or by presenting his monthly valuation, the majority of all routine supervision is carried out by the team members acting on their own initiative. The resident engineer can establish procedures for regular inspections and encourage his staff to maintain a close watch on the contractor, but in the end he must rely on the enthusiasm and diligence of his team. The members work in close contact and know each other personally; to be effective, they must cover and support one another without rigid lines of demarcation. This dependence on people, both as individuals and as a group, is the outstanding feature of the site team and can be its great strength or its fundamental weakness. In his role as its leader, the resident engineer has the vital responsibility of ensuring that the right working environment exists in which his team can be a success.

Staffing a contract

Figure 5 shows a typical organisational chart for a site team, in

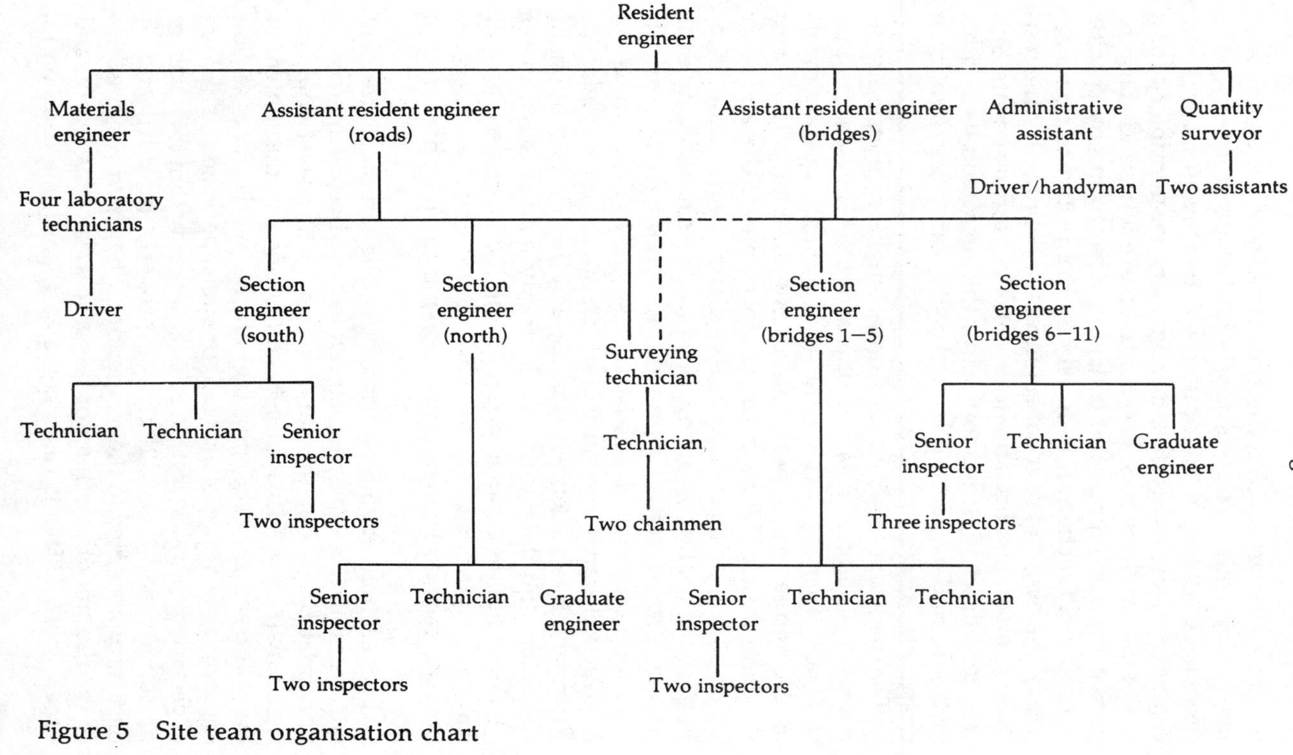

Figure 5 Site team organisation chart

this case one supervising the construction of a medium sized motorway with a large proportion of structural work. The resident engineer is supported by two assistant resident engineers each having 'line management' responsibility for a major part of the works. Few contracts are sufficiently large to justify the appointment of a deputy, free from any specific operational function. It is usually more effective to nominate one of the assistant resident engineers to deputise for the resident engineer in his absence. Working arrangements and responsibilities must be notified to the agent and he and his staff will find it useful to have the site team's organisational chart.

The assistant resident engineers' main responsibility is managing the team's day to day activity. They ensure that staff are properly deployed, records kept, measurements agreed and recorded, deal with routine communications with the contractor and liaise with statutory undertakers and other 'third parties', supervise any new design work or modifications to the contract drawings produced on site and handle the normal procedures of approval and acceptance of work. It must be noted that the assistant resident engineers have no contractual status beyond that of 'assistants to the Engineer's Representative' as defined in clause 2 of the ICE Conditions of Contract, and their authority is strictly limited (see Chapter 3). Although the resident engineer will look to his senior staff for their informed advice when exercising his delegated powers, each decision or instruction must remain his personal act.

The section engineers report to the assistant resident engineers. In this example their duties have been assigned on a territorial basis, but the sub-division at this level could be functional, the choice is largely conditioned by the content of the contract and the phasing of the work. The section engineers are at the 'sharp end' of site supervision. They are expected to know everything that goes on within their sphere of activity and ensure that it is properly recorded and, if necessary, reported to their superiors. They have responsibility for the smooth running of the processes of inspection and supervision. Difficult cases come to them for a decision or referral up to the assistant resident engineer and to a large extent they set the standards for approval and acceptance.

Each section engineer heads a group of engineers and/or technicians and inspectors, the mixture will depend upon the type of work to be supervised. Engineers/technicians generally concentrate on 'technical' activities such as checking calculations and setting-out, doing survey work and assessing performance; inspectors (or clerks of works) are involved mainly in the supervision of workmanship and the 'method' aspects of the specification, and are usually recruited from skilled craftsmen such as carpenters or bricklayers. These divisions cannot be strictly observed, however, for every member of the section engineer's group shares in the continuing task of supervision, although each may develop a particular specialisation.

There are 'true' specialists amongst the site staff. The quantity surveyor (or measurement engineer) deals with the valuation of the contractor's work (see Chapter 11). Under the ICE Conditions of Contract he is not an official of the contract, has no specific duties or powers and serves in a supporting and advisory role — albeit an expert one. As a member of the site team, the quantity surveyor must observe the Engineer's obligation to 'act fairly as between the parties'.

The materials engineer is often supplied, along with the other laboratory personnel, by one of the specialist testing consultancies. The appointment is usually made by the Engineer, with the employer's approval, and the post forms part of the site team. His task is to undertake or arrange the testing (on site or off) of the materials and components being incorporated into the works, to report the results to the resident engineer and advise upon their implications. The professional independence and competence of the materials engineer is a great asset to the resident engineer for it encourages a confident approach to the problems of acceptance and approval and, more importantly, earns the resident engineer's laboratory a reputation for impartiality and accuracy which can be invaluable in any dispute.

Administrative assistant is a cumbersome title, necessitated by the great range of duties undertaken by this team member:

secretary to the resident engineer, typist, filing clerk, telephonist, radio operator, receptionist, office manager. It is easy to underestimate the contribution which effective communications and efficient office procedures make to a project, for when these matters are being dealt with smoothly they attract little attention. Nevertheless, the chaos which results from unanswered telephone and radio calls, misplaced messages or an inadequate filing system has a significant effect on the performance of the site team, and the time absorbed in carrying out these tasks properly is such that to leave them to the engineering staff is to misuse resources on a grand scale.

On a big site the staff will be assisted by chainmen and perhaps a driver/handyman, normally supplied by the contractor and paid for under billed items. These jobs are important and the resident engineer should make sure that the agent recruits people who will enjoy the work, fit in well with the team and stay for the duration of the contract. Time spent on achieving this aim will be repaid many times over.

The resident engineer's main task is to provide the leadership which will make this team effective. Volumes could be written on the qualities and skills which he will need, and there are a number of excellent books on the general topic of man-management which a resident engineer might profitably read, but the basic guidelines can be simply expressed. A good resident engineer will be unmistakably in charge of his team whilst recognising that he cannot be an expert on everything; he will take his personal responsibility for all that the team does seriously without trying to do every job himself; he will keep everybody — himself included — fully informed about what is going on around the site but avoid stifling initiative with too many instructions. Above all, he will work hard to ensure that each member of the team feels useful and busy: a task in which he will not fail for lack of raw material!

There is no such thing as a 'typical contract' and so to list a series of standard tasks or a daily routine for engineers, technicians and inspectors would be unrealistic. However, some impression of the range of work covered by these members of the resident engineer's staff can be obtained from a review of the main duties assigned to their teams by two of the section engineers in the site organisation illustrated in Figure 5. The contract is for the construction of a motorway; the period is a week in the first summer of the two-year project.

Roads team (north)

(1) Graduate engineer: organise the collection and processing of data on the output of the earthmoving fleet; arrange with the Department of Transport for temporary traffic signals where the haul route crosses a trunk road; attend (with the assistant resident engineer) a meeting with the contractor and statutory undertakers to discuss major diversions, and then write up the minutes; routine earthworks measurement (with one of the site quantity surveyors) for the monthly certificate; joint inspection with adjacent landowners of newly-completed fencing, and agree remedials with the contractor.
(2) Technician: routine levelling (jointly with the contractor) on the earthworks for the monthly measure; 'spot levelling' the cuts and fills as part of the output measurement exercise; record positions of soil samples taken by the materials engineer; accompany photographer on monthly visit; local checks on the contractor's setting-out for earthworks and advanced drainage; check line and level of completed pipe run for approval.
(3) Senior inspector: record deployment of plant and labour (with particular attention to the earthworks operation); random checks on round-trip times for motorscrapers and average loads; check condition of haul routes and road crossings; inspect and assess (with section engineer) the extent of unsuitable material exposed in main cutting; general supervision of earthmoving.
(4) Inspector: supervision of excavation operations and advanced drainage works; check surcharge on cutting floor; inspect temporary drainage outlet; inspect trimmed side slopes for approval.
(5) Inspector: close supervision of compaction; check surcharge on fill, and arrangements for disposal of surface water; record round-trip times for motorscrapers.

Structure team (bridges 1 — 5)

(1) Graduate engineer: assist assistant resident engineer in checking contractor's falsework proposals; supervise loading test on trial pile; routine measurement (with one of the quantity surveyors) for the monthly certificate; visit (with section engineer) fabricating yard to inspect steel footbridge under construction; prepare variation order for incorrectly detailed reinforcement at bridge 1.
(2) Technician: inspect steel and formwork for approval (bridge 2 — south abutment, bridge 4 — base); arrange with British Rail for flagmen and lookouts at railway bridge site; local check on setting-out for piling at bridge 3; attend (with laboratory staff) loading test on trial pile.
(3) Senior inspector: supervise concrete pours at bridge 4 (north and south bases); inspect reinforcement cages for piles being fabricated on site; general supervision of steelfixing and formwork in progress at bridges 1, 2 and 4.

(4) Inspector: inspect steel and formwork at bridge 2 (south abutment) and supervise concrete pour; supervise steelfixing and formwork and progress at bridge 2 (north abutment).
(5) Inspector: inspect steel and formwork at bridge 4 (base) and supervise concrete pour (with senior inspector); supervise striking of formwork and backfilling of over-excavation; visit precasting yard (with section engineer) to inspect bridge beams.
(6) Inspector: supervise steelfixing and formwork at bridge 1 (wing walls); supervise backfilling at (south abutment); assist in supervision of concrete pour at bridge 2.

The contractor's site organisation is less likely to follow a typical pattern than that of the resident engineer. The extent of sub-letting, the amount of other work which the firm may have in the locality and the relative importance of the scheme are just three of the many variables which can influence the arrangements for staffing a particular site. The agent and his sub-agents correspond in broad terms to the resident engineer and his assistant resident engineers but there is no equivalent in the resident engineer's team to the third major figure on the contractor's staff, the general foreman. The 'g.f.' has the vital task of controlling the day to day activities of the plant and labour on the site. This is achieved through the various foremen and gangers who report to him and, very often, by what may be politely described as a highly personal and interventionist style of man-management. Most general foremen are blessed with strong, not to say eccentric, personalities and considerable dramatic talents. All of them are at pains to point out that they have 'seen it all' - some of them actually have.

Setting-out is a critical factor in the progress of site operations: this is reflected in the large numbers of 'engineers' (a traditional description which has nothing to do with professional status) who work under the direction of the sub-agents on every aspect of this activity, from establishing master survey stations to nailing up the profiles showing the gradients of side slopes. This group will vary in strength as the amount of setting-out work fluctuates. Self-employed or agency-supplied 'engineers' are often brought on to the site to supplement the contractor's own staff, but usually they have no authority to direct plant and labour or to receive instructions from members of the resident engineer's team.

There are other differences of function and emphasis. Few contractors set up a fully-equipped laboratory, although there may be a small testing facility on the site or at a nearby supplier's yard and a member of the agent's staff might be given responsibility for the sampling and quality control of materials. The numbers of quantity surveyors, and the forcefulness of their presence, varies from company to company, as does the policy of employing these specialists direct or through consultants. The agent's office deals with the ordering of materials and fuel, the stocking and issue of stores, the processing of bills and accounts and, most important of all, the feeding and paying of the work-force so there are, therefore, many more ancillary staff than are found on the resident engineer's team. These differences mean that there can be no direct comparison between the size of each team, nor is there any straightforward relationship between the value of the contract and the number of staff needed to supervise it. When the Engineer makes his recommendation to the employer on the composition of the site team he will take into account the complexity and pace of the work, the extent and nature of the site, the likely amount of involvement with subcontractors and outside organisations and the probable level of sampling and testing. Subjective assessment of the particular contractor's ability and the need for close supervision must also come into the equation.

It is often said that numbers are no substitute for site experience, and although this quality is of great value there are good reasons why the site staff should not be entirely composed of veterans. The design process suffers if there is no direct interchange of information between drawing office and site, and the best way of achieving this is to ensure that some designers are given the opportunity to work on site and see at first hand the methods of working and the practical problems of supervisors and contractors. Furthermore, a construction project is too valuable a training ground not to be used, and the resident engineer's team ought to contain a proportion of staff who are there to gain experience.

A wealth of site experience is no guarantee of an effective supervisor. Just as important is what may be termed 'site ability', a mixture of enthusiasm, knowledge, curiosity, initiative, methodical working habits and sheer dogged persistence which comes as a natural quality rather than a qualification. This need for an amalgam of talent and training has been recognised since the earliest days of the engineering profession, although some other requirements of those times have diminished in importance. Joseph Bennett, Brunel's chief clerk, wrote in 1853 that any prospective staff member should be:

'... of gentlemanly habits as well as of good and gentlemanly connections ... and either by his special education and his tastes or by his natural turn and ability he shall have given sufficient reason to suppose he will succeed in the profession ...'.

Working relationships

The resident engineer and the agent are working towards the same end, the satisfactory completion of the project, but they are judged by different standards and have different priorities. Inevitably there will be disagreements over the means by which their shared goal is achieved, but their relationship must remain professionally correct. This is the final level at which site decisions are made and disputes settled. Any breakdown in communications could paralyse the whole contract.

Both the resident engineer and the agent have a personal responsibility for establishing and maintaining a constructive relationship between themselves and their teams. Each of them, however, belongs to an organisation which extends beyond the site and which, if it does not recognise their special positions, can undermine their best efforts. The most valuable contribution to the success of the contract which the Engineer's and the contractor's head offices can make is to ensure that, wherever possible, site decisions are kept at site level. Provided a matter lies within the resident engineer's delegated powers, or the less formal limits on the agent's authority, there should be no interference except through the machinery of the contract. The ICE Conditions of Contract (clause 2) allow the contractor to refer any decision made on the site to the Engineer and call upon him to 'confirm, reverse or vary' it. This provision is available to a contractor who wishes to obtain a review of a decision without declaring a formal dispute and, as long as it is not used to harass the site staff, is a worthwhile safeguard. What must be deprecated is the contracts manager or a director telephoning the Engineer or one of his senior staff at head office to put 'unofficial' pressure on the resident engineer. Contractors are subject to commercial forces and it is understandable that sometimes this approach will be tried. There is, however, no reason why the Engineer should deal with matters in this way, unless he has no confidence in his representative on site, in which case the resident engineer should be replaced, not by-passed. The existence of any informal appeal against the resident engineer's decisions not only undermines his standing, but also subverts the whole contractual process.

There are occasions when it is reasonable to take a flexible attitude to the specification or other requirements of the contract documents, but decisions of this kind are reached through a thorough knowledge of all the relevant facts, a certain amount of prediction, a large measure of engineering judgement and a sound understanding of the 'tactics' of the contract as a whole. Compromises are best left to the man on the spot, acting within his authority and using his first-hand experience: head office decisions should follow the letter of the contract, harsh though it may be.

The site team works effectively only when the chain of command carries the resident engineer's authority down to the decisions of the most junior member. This is achieved in two ways. Firstly, decisions must be consistent, so that the agent and his staff know that there is no point in seeking out different members of the site team to get different decisions until they finally obtain the one they want (the contractor's men will still try, of course, if only out of habit). If the resident engineer's general approach to supervision is fully explained to his staff and the reasoning behind specific decisions is made known, then any member of the team faced with a particular question can make a reasoned assessment of what the resident engineer himself would have decided. The best compliment an agent can give to a site team is to protest, 'It doesn't matter which one of you lot we ask, we always get the same answer'.

The second way in which the team's authority is consolidated is by giving the members confidence that their decisions will stand. To a large extent this results from the development of a consistent 'team view', but complete uniformity can never be achieved and other means must also be used to foster the self-assurance of each member. There will inevitably be times when an engineer, technician or inspector is asked for a decision and does not choose the option the resident engineer would have taken. If the decision is contractually wrong it must be reversed whether it is to the disadvantage of the contractor as, for example, when some unsuitable material has been mistakenly accepted as backfill, or to his benefit such as when, for instance, a concrete surface has been rejected by an inspector who is seeking a standard of finish higher than that laid down in the contract documents. In many cases, however, it is not that the decision is wrong, but that it is not the best one in the circumstances: unnecessarily expensive stone may have been used to fill a soft area, or old materials accepted as formwork when new timber could have been insisted upon. The resident engineer has to consider two things, the future standing of that team member and the effect on the permanent works. More often than not the wisest course is to back the decision and uphold the authority of his staff, as individuals and collectively, whilst explaining to the person concerned why a different response ought to have been made. There may be problems with an irate agent or the employer may have to pay a little more than might have been necessary but, in the long run, the value of a confident, authoritative team far outweighs the costs of a few isolated errors of judgement. This is not the resident engineer blindly saying 'my staff, right or wrong', it is a demonstration of the unqualified support that is available to each team member whenever it is deserved.

The resident engineer must give his staff the opportunity to undertake their tasks with the minimum of interference, he must rely on their judgement and show he does so by resisting the temptation to act as a superior kind of inspector. Every team member has the right to be spared two things: having one of their decisions reversed in front of the contractor; and arriving at a part of the works to find that the resident engineer has passed by and already approved, or rejected, it. If the resident engineer sees a mistake, then the person responsible should be told and allowed to go back and correct it himself and any requests to inspect work should always be referred to the appropriate personnel except in genuine cases of urgency.

It would be a strange contract where the resident engineer's team and the agent's staff had no disagreements. They are not, or should not be, antagonists, but they are nevertheless adversaries in the sense that they may sometimes find themselves serving different interests and so pulling in different directions. As long as the pulling resolves into reasonable forward motion (and every engineer knows about the parallelogram of forces) then the contract is progressing, and it is one of the main tasks of the resident engineer and agent to ensure this.

It is inevitable, given the pace of construction work and the immediacy of the problems it generates, that relationships will be robust, especially at assistant resident engineer and subagent level. This is the 'court of first instance' where significant disputes, which have probably already caused a good deal of friction, are brought for resolution. Assistant resident engineers are usually chosen for their 'R.E. potential', and most want to show their effectiveness by dispensing wherever possible with the assistance of their superior; sub-agents have a reputation as the 'hard men' of any contractor's organisation and often wish to live up to it. Both are understandably keen to advance their careers. Wise resident engineers and agents will see the rough and tumble of this competition as a healthy safety valve for the site, and a very useful buffer to their own relationship which, as the corresponding 'court of appeal', has to remain on a businesslike, unemotional level.

Good communications require not only speed and clarity but also definite channels through which information and instructions can flow. The contractor's staff must direct requests for information or for approval of completed work to the appropriate member of the resident engineer's team, either directly or by leaving a clear message. For his part, the resident engineer must ensure that team members can be contacted easily or, if absent from the site, have their duties temporarily reassigned. It is essential that 'official' contact is limited to the agent's immediate staff (and this includes foremen and gangers controlling operations on the site) and that instructions or other contractual directions are never given to labourers, machine operators or other members of the work-force. It is quite improper for site staff to deal directly with employees of the contractor who are in no position to judge the effect or importance of what they are being told, particularly when, as is often the case, extra work, delay or other consequences likely to result in extra costs are involved. The agent has a right to expect, and will insist upon, all communications affecting the running of the contract to go through him or those members of his staff who can judge their importance. If there is any doubt about the agent's chain of command, if for example, the system works badly or slowly causing abortive work or expensive delays the contractor must put it right and bear the cost—it is not for the resident engineer and his team to develop 'short-cuts'.

Most experienced site staff will acknowledge the contractual propriety of this description of site communications, whilst admitting that there are some notable differences between the theory and its practical application. On large sites it may prove difficult to contact individuals immediately or get them quickly to a particular location, especially when several operations are going on simultaneously. In these circumstances, a supervisor who sees unacceptable workmanship or materials being incorporated into the works is faced with a difficult choice. He may recognise that the operation could easily be stopped if only someone in authority on the agent's staff could be contacted, but knows that if left for an hour, or until tomorrow, it will be disruptive and expensive to carry out proper remedial work. (If he is of a suspicious nature, he may even suspect that this thought has crossed the agent's mind and may explain the disappearance of most of the agent's staff.) To return to the site offices and send a letter across to the agent would show an unimaginative and pedantic view of the business of supervision and would do nothing to advance the employer's interests. The rule-of-thumb for supervisory staff dealing with the realities of site operations is this: when work is going on which will have to be rejected, inform whoever is involved that it should stop and confirm it with the agent's staff as soon as they can be contacted. If additional, new or changed work is required, issue the instruction only to the agent or those of his staff who have authority to receive it. The successful operation of this rule depends to a great extent on the respect, albeit well hidden, which the contractor's men have for the resident engineer's team.

The working relationship also depends on the recognition by the resident engineer and his team of the very real difficulties under which the contractor has to operate. The separation of career paths which seems to be a permanent feature of the engineering profession and the construction industry (these two phrases illustrate the division) means that few of the resident engineer's team, with the possible exception of the inspectors, have any direct experience of contracting. Lacking that background, they may fail to recognise the organisational problems of the agent and his staff: deploying plant and labour to meet different needs in different parts of the site; getting the right materials delivered in the right place at the right time; co-ordinating sub-contractors' activities; all of this and more, and subject to the inevitable intervention of the weather. For many engineers the relatively straightforward exercise of planning and constructing an extension to their home has proved a chastening experience from which they have emerged as more sympathetic and pragmatic site supervisors, having seen in miniature just some of the frustrations of civil engineering construction.

Nevertheless, it is essential that the employer's agreement with the contractor is not compromised: he is not merely buying materials and hiring plant and labour, but purchasing expertise, experience and organisational ability. Furthermore, the bargain which has been struck places certain obligations, with their attendant commercial risks, firmly with the contractor and consequently the problems of stretching resources to meet tight programmes, of securing uninterrupted and efficient materials supplies, of controlling sub-contractors and even of dealing with the British climate have to be overcome by the agent without any loss of quality or unnecessary extra cost to the employer. It is the resident engineer and his team who have the continuous task of striking the right balance between working *with* the contractor and *for* the employer, whilst always maintaining their independence of judgement. Their best, indeed, their only, policy is one which requires effort and application but can be simply expressed: firm but fair.

Sub-contractors

Few contractors nowadays can afford to keep within their own organisations all the skills and resources necessary to carry out every operation on a major civil engineering project. The solution to this problem is to sub-let, with the Engineer's approval, certain parts of the works to other firms who act as sub-contractors under agreements with the main contractor. In most cases the Federation of Civil Engineering Contractors' standard form of sub-contract (the 'blue form') is used, but it is not unusual for modifications to be made depending upon the relative bargaining strengths of the main contractor and sub-contractor. The extent of the sub-contract can vary enormously. A small, self-contained and technically-difficult operation, such as the post-tensioning of a structure, may be sub-let to a firm which provides expert staff with their own specialised plant. On the other hand a complete part of the works, the drainage, for instance, may be let out with everything from equipment to skilled craftsmen and foremen supplied by the sub-contractor. The presence of a 'labour-only' sub-contractor on site supplying gangs of men to work on various operations for which the materials, plant and foremen are all supplied by the main contractor does not amount to sub-letting under the terms of the ICE Conditions of Contract.

Clause 4 of the ICE Conditions of Contract makes the main

contractors he may employ 'as fully as if they were the acts ... of his agents, servants or workmen'. Even when the employer specifies in the contract documents that a particular firm is to be used to supply certain materials or execute some part of the works (a 'nominated sub-contractor'), the workmanship, programming, safety precautions — and mistakes — of that firm are covered by the main contractor's obligations just as much as if he chose the sub-contractor himself or did the work with his own resources. The reasons for this are both contractual and practical. In legal terms, the only agreement to which the employer is a party is the one he has with the main contractor. There is what is known as 'privity of contract' between them and each can enforce the terms of the agreement against the other. There is no 'privity' between the employer and any sub-contractor because they have made no agreement together, and so it is by ensuring that the main contractor takes full responsibility for all the works that the employer retains his right of action for any default or defect. On a practical level, it would be very difficult for the resident engineer to supervise several sub-contractors, each subject to a different agreement with the main contractor (possibly including private or informal arrangements associated with other work off-site), and each controlled to a greater or lesser degree by the agent.

The most important consequence in terms of site communications is that the resident engineer's staff do not deal directly with sub-contractors. All contact must be through the agent or his staff. Once again, this is a statement of theory. In practice, many agents, with their staff spread thinly over the site, will encourage direct contact on routine matters provided the proper procedure is followed on all items of a substantial or potentially controversial nature. The rule of thumb for dealing with the main contractors can be applied to the sub-contractors: if work in progress is seen to be unacceptable then it is sensible to advise (*not* instruct) the sub-contractor to stop and avoid any further abortive effort whilst the agent's staff are being contacted. If a positive instruction, involving additional, changed or completely new works, is to be issued then this must go through the main contractor. The agent must not only know what the resident engineer and his team are saying about the sub-contractor's operations, he must also be allowed to decide for himself whether any varied or new work will be given to a particular sub-contractor and if so, on what terms. This is not just a matter of etiquette. Very serious issues of liability for costs and the control of operations are involved. Writing a term into the contract which requires the agent to assume full responsibility for sub-contractors is an empty gesture if he is then by-passed (or allows himself to be by-passed) on the site.

The informal guidelines described above apply to the routine of supervision, but if there is any doubt, or if contractual problems can be foreseen, then the following formalities must be observed:

(1) Always deal with the agent or his staff (preferably through a specific contact — a sub-agent, perhaps — nominated when the sub-contractor is approved).
(2) Always confirm discussions, telephone conversations or spoken messages in writing.
(3) Never discuss contractual matters with a sub-contractor if the agent is not represented.
(4) Never address any communication to a sub-contractor.

The only exception is when the action or inaction of a sub-contractor causes a risk of personal injury. In such an emergency it would be right, indeed, there would be an obligation, for any member of the resident engineer's team to approach a sub-contractor directly, although it would still be necessary to involve the agent as soon as possible.

Letter-writing

Throughout the contract the resident engineer and the agent, personally and through their respective staffs, engage in a continuous dialogue. Opinions, proposals, approvals, rejections, criticisms, warnings, decisions, instructions, explanations and requests for information will pass to and fro in a steady stream. Putting them in writing ensures that they are preserved for future reference and are placed clearly before the recipient in such a way that they cannot be ignored. The spoken word can easily conceal an ambiguity or contradiction and is always open to the interpretation of the listener. Cold print conveys only what is on the page, and whilst interpretation and 'reading between the lines' is always possible, the scope for misunderstanding, wilful or accidental, is limited.

The written record of communications between the site team and the contractor is therefore of value both during the progress of the contract and in any subsequent review when it may come before persons who were never present on the site. They may have to read and absorb the contents of many items so it helps if everything is neatly and uniformly presented and,

A329 AMEN CORNER TO DOWNSHIRE WAY

FRENCH KIER

Location DONCASTLE SLIP ROAD Date 4.6.82

CH. 0-15 (BOTH SIDES)

Approval is requested for SETTING-OUT PINS FOR KERB LINE AND CHANNEL – LINE AND LEVEL

Ready at 3:00 pm.

Submitted by M.J. Skinner

BERKSHIRE COUNTY COUNCIL (R.E. OFFICE)

Received by B.C.C. Time 2.50 pm Date 4 June '82.

Approved/~~Rejected~~

Reason for rejection OK except ch 90 (west) – 100 mm low

Signed P. B. Camden
3.40 pm.

Figure 6 Typical form for approval of setting-out

preferably, typewritten. It is not necessary for every communication to be written as an individual letter, and in the case of routine, recurring correspondence, such as requests for the approval of setting-out or for work to be covered up, the requirements of permanence and clarity can be fulfilled just as well through the use of pre-printed forms. (Figure 6 gives a typical example of such a form.)

Staff are sometimes discouraged at the thought of expressing themselves in print, particularly when complicated sequences of events or long and detailed discussions have to be recorded. Nevertheless, the resident engineer should make it his policy that team members prepare their own correspondence and, although every letter must go out under his signature, he should restrict corrections and modifications to those necessary for the proper running of the contract. Site staff do not need the wit of an Evelyn Waugh nor a Churchillean mastery of language to be effective letter writers, indeed, humour is entirely misplaced in contract correspondence and content is far more important than style. The object is to transmit relevant statements or opinions as clearly as possible and to allow the recipient to consider them and respond accordingly. If the result is agreement, then it is on record, if dispute, then the boundaries of the disagreement are established.

The site files of any large contract will contain numerous lessons on letter writing.

'I refer to your letter dated 21 February, but received on 7 March ....'
Correspondence should never be back-dated, even when there are 'good' reasons, such as staff absences, for the delay. Any excessive delay in receipt should be recorded. If a message is urgent, send it by any reasonable means, handwritten on scrap paper if nothing else is available, and confirm it with a typewritten letter later.

'Further to today's joint inspection of the newly laid area of floor slab, it was agreed ....'
Always give full descriptions, particularly of locations. The writer and recipient may know exactly what is being referred to, but others without their intimate knowledge of the works may have to read, and understand, the correspondence later.

'I have discussed this matter with members of your staff over the last few days, but cannot accept that ....'
Be specific about names and dates. Keep to the convention that 'I' refers only to the resident engineer (whose signature is on the letter), 'you' means the agent (to whom the letter is addressed) and everyone else is 'my section engineer Mr. Jones', 'your assistant Mr. Smith' and so on.

'I think the colour difference in the concrete will still be visible after weathering and further remedial work may have to be carried out.'
Be positive. The agent will do nothing on the strength of a vague admonition about which the resident engineer appears to have no strong feelings. It would be better to say:
'The colour difference is unacceptable and I consider weathering will not eliminate it, although I am prepared to delay my final inspection for three months at your request.'

'I enclose laboratory results on the imported fill which is not in accordance with the specification and await your early observations.'
Do not prolong correspondence unnecessarily. If this material is to be rejected and removed then the letter should say so. Remember that the agent's observations will be made from his point of view, which is likely to differ significantly from the resident engineer's. Ask for them only when they are really needed.

'The gravelly clay from the north end of the cutting should be used as it is a more suitable material than the silty clay you propose to employ.'
Choose the words carefully. This writer is going to regret introducing the phrase 'more suitable', for the agent will point to the normal earthworks classification of 'suitable' and 'unsuitable', ask for a definition of the new term and suggest that its use really means that some 'suitable' is wrongly classified and is in fact 'unsuitable'.

'The testing to which you refer is covered by the contract and cannot be paid for as an extra ....'
Be precise on contractual matters. If the testing is covered by a bill of quantities item, then details should be given. If the writer is relying on the general obligation in clause 36 of the ICE Conditions of Contract to make allowance for reasonable testing, then this should be stated. It will save time and confusion later.

'... as a consequence of these incorrect ground levels it will be necessary to extend the 300m french drain a further 50m ....'
Do not casually insert a variation of the works into a letter. The agent is entitled to have every change formally notified as an 'Ordered Variation', (see Chapter 10). Anything less is sloppy contract procedure, and will cost somebody money.

'Your request for a three-week extension is rejected.'
Never go outside the limits of delegation (extensions can only be decided by the Engineer). If a situation arises which requires the involvement of the Engineer, details should be passed to him as soon as possible and the agent informed accordingly.

'... or else I will suspend the works.'
Never bluff or make empty threats - the agent may invite the resident engineer to carry them out.

Site correspondence is meant to be business-like and so will be short on pleasantries. The resident engineer 'instructs' the agent, to remove unsuitable materials, perhaps, or to do additional work, and does not use the word 'please'. This is not deliberate rudeness but the conventions of formal letter-writing. The constant theme running through the letter files will be criticism: of the contractor's methods, workmanship, materials and performance. This is an inevitable consequence of the resident engineer's supervisory function. The contractor's good work does not pass unnoticed, nor even unrecorded, for every certified payment on a bill item stands as proof of the Engineer's 'satisfaction', but it is in the interest of all concerned that every alleged default is dealt with formally and in detail so that the final settlement can be made fairly.

If the resident engineer and his team follow this principle conscientiously the files will probably convey to an outsider the impression of a struggle, conducted with a grim and unnecessary attention to detail. Anyone experienced in contracting will know that the correspondence represents just one aspect of what is usually an amicable working relationship and will not be misled. Occasionally, if a dispute continues after the completion of the works, the contractor's senior management may suggest that the tone of the letters indicates a 'breakdown in communications' and shows the 'unreasonable attitude' of the site staff. The resident engineer's team need not be concerned by suggestions of this kind provided the letters are kept factually accurate, any opinions are clearly identified as such and extreme language is strictly excluded: supervisory staff are sometimes 'concerned' or 'dissatisfied', but never 'horrified', 'amazed' or 'disgusted'.

It is impossible to predict during a project what matters will have assumed importance by the time it has ended, or what records will be relevant to the final settlement of the contract. Reading the files at the end of a job is an instructive exercise, for it invariably illustrates one of the paradoxes of site supervision. There will be numerous letters, all of which seemed important at the time, referring at length to matters now long-forgotten and insignificant, and there will be gaps where a single short letter, the one somebody meant to write but then decided was unnecessary, would be enough to settle a disputed measurement or fix the liability of an extra cost. There is no substitute for agreed, comprehensive records. The letters provide a running commentary, with corrections and observations, on the progress of the works which is contemporary and to which both the resident engineer and the agent have been free to contribute. Its value is enormous. If in doubt—put it in print.

Meetings

Throughout the life of the contract meetings will be taking place on the works, at the site offices, and at supplier's yards. In many cases they will be arranged at short notice and may even develop unplanned out of some incident on site, others will be formally called, or held at regular intervals. The main object of any meeting is to come to a decision, even if it is a decision to disagree, and record it, together with any relevant information.

Informal meetings usually result in nothing more than an agreement to remove some unacceptable work or to continue with a particular method of operating, decisions made within the requirements of the specification or the drawings. A note in the participant's diary is sufficient in such circumstances. If, however, there is any dispute over facts or liability, then the resident engineer must notify the agent in writing of the time and place of the meeting, the names of those present and the substance of their discussion. Meetings with sub-contractors in the absence of the main contractor have no contractual status, they are no more than conversation.

From time to time, it may be necessary to convene a formal meeting to discuss a particular subject which has become important to the progress of the contract. Meetings of this kind usually fall into one of three categories: discussions between senior members of the site organisations, often involving the resident engineer and the agent personally; meetings with the main contractor at which a sub-contractor is present; meetings involving a 'third party' (a statutory undertaker, for example) at which the agent may or may not be represented.

Some general principles apply to all formal meetings. Whether requested by the resident engineer or the agent, the meeting should be called by the resident engineer and he, or one of his senior staff, should act as chairman because, as the engineer's representative, he stands independent of the two parties to the contract. An agenda is very useful when a meeting is dealing with more than one topic. It gives notice of the matters to be discussed, so that the participants can come prepared, and it provides a means of focusing attention on the business at hand. The resident engineer should arrange for someone to take notes so that a record of the discussion and any conclusions can be issued promptly. Formal minutes are sometimes required, especially when 'third parties' take part, but it is usually sufficient to summarise the main points. The minutes or notes need not set down the words actually spoken, provided they convey accurately the sense of what has been said. What they must not do is introduce new material, omit significant points or, worst of all, put forward someone's idea of what *should* have been said.

When meetings are held between the staffs of the resident engineer and the agent, formality is at its lowest level and there is a tendency to range over many topics, only some of which may be related to the original purpose of the discussion. Open exchanges of opinion between the staffs of the resident engineer and the agent are to be encouraged, but all concerned should take care to avoid making hurried or ill-advised decisions, and the decisions that are taken should be recorded properly however informal the atmosphere. Meetings without a proper record, like letters which never get written, have a disconcerting habit of becoming important after the event.

Because of the specialised nature of their operations, sub-contractors may appear at meetings along with the main contractor. Provided their presence is at the suggestion of the main contractor, the resident engineer should have no reason to object, indeed, their attendance is likely to improve communications. Nevertheless, it must be made clear at the outset that any direct discussion between the resident engineer and the sub-contractor takes place only in the presence of the main contractor and only because it contributes to a better understanding of the points at issue. The main contractor must not be allowed to stand aside and leave matters to be resolved by the resident engineer and the sub-contractor: responsibilities must always rest exactly where they are placed by the contract. Meetings with 'third parties' share, to some extent, the same pitfalls. The resident engineer must guard against any blurring in the limits of responsibility, ensuring that his attendance is not mistaken as an offer to accept duties which belong elsewhere. The detailed programming of public utilities' diversions, for instance, is generally the contractor's obligation. Because of their involvement with these organisations at the planning stage, the resident engineer's staff are usually willing to assist in organising meetings and will attend them because of their importance, but the agent must be clear that he must arrange and co-ordinate these activities.

One meeting deserves special attention. On most contracts a regular progress meeting is held, usually at monthly intervals, and acts as a valuable reference point from which a continuous survey of the whole of the project can be maintained. For the meeting to be used to full advantage, it should be chaired by the Engineer or a senior member of the head office staff and the contractor should be invited to send a representative of similar status. It provides a forum in which the contract can be discussed at a high level, gives an early warning of disputes which may not be capable of resolution at site level and ensures that any legitimate complaints against the supervisory team are not stifled at source. A formal agenda is essential, and is best prepared around a series of standard main headings (the major divisions of the bill of quantities, for instance). Under the main headings sub-headings define the

NAME OF CONTRACT

Progress Meeting No. 11, to be held at the Resident Engineer's offices, 2.00 pm on Thursday October 7th.

A G E N D A

1. Minutes of last meeting

2. General Progress

   2.1. Progress Review for September
   2.2. Interim Assessment of Extension

3. Preliminaries

   3.1. Temporary traffic signals at scraper crossing
   3.2. Diversion at Bridge 5.

4. Site Clearance, Fencing

   No items

5. Drainage

   5.1. Delayed completion of outfall at Upper Farm
   5.2. Supplies of porous pipes

6. Earthworks

   6.1. Contractor's proposals for the remainder of the season and for protecting the earthworks over the winter
   6.2. Contractor's request for information: variations to capping layer

   etc.

12. Accommodation Works

   No Items

13. Statutory Undertakers

   13.1. Diversion of 11kV overhead line
   13.2. Programming of Gas Board works

14. Landowners, Insurance Claims Etc.

   14.1. New claims notified to Contractor
   14.2. Encroachment on land at Mill Cottage

15. Safety

   15.1. Illumination of warning signs at Northern Roundabout
   15.2. Trench excavations and means of access for inspection

16. Any other Business

   16.1. Proposed visit of Society of Civil Engineering Technicians

17. Date of Next Meeting

Figure 7 Extracts from progress meeting agenda

business of the particular meeting. Figure 7 shows the agenda for a typical progress meeting on a large road contract. The meeting should be properly minuted, as the authority of those attending and the importance of the subjects discussed give the record of the progress meetings a very high standing amongst the documentation of the contract.

The style and content of a formal minute are best illustrated by the following example.

2.1 General Progress
Mr. Edwards referred to the exceptionally heavy rainfall in March which had disrupted bridge construction and delayed the start of earthmoving. The R.E. said that the rainfall figures kept by the site laboratory would be compared with the Met. Office's local records for the last ten years; he referred to his comments at earlier progress meetings about the optimistic start date for earthworks shown in the contractor's programme. Mr. Brown asked whether the contractor was seeking an extension of time under clause 44. Mr. Edwards said that this was under consideration. The R.E. said that details of any claim for an extension should be submitted promptly, as the period of heavy rain had been three weeks ago and further delay was unnecessary. The agent agreed to present all the available information within a week.

A properly-convened meeting, called at reasonable notice, offers an ideal opportunity for getting through a lot of business efficiently, provided those attending have prepared for it, understand what is to be discussed and know what they want to get out of the meeting. This may seem a statement of the obvious, but it is nevertheless true that there is often no 'pre-meeting', and it is not unusual for staff to meet the contractor's representatives not knowing the resident engineer's view on the subject under consideration and with no specific directions on what matters they should discuss, or whether they should speak at all. If a meeting is worth staff time in attendance, it should also be worth time in preparation.

'Third parties', the public and the press

'Third parties' have an interest in the execution of the works but are not part of the agreement between the employer and the contractor. Typical members of this category are statutory undertakers, British Rail, and the landowners and occupiers whose property is affected by the Works. Any person or organisation can become involved as a 'third party' if they suffer loss, damage or injury as a result of the execution of the contract. This includes the motorist who damages his car on some item of plant left unlit on the highway at night, as well as the owner-occupier who claims his home has been damaged by vibration from piling operations. In all cases, the 'third party' has some right which is affected by the construction of the works. It may be a right conferred by legislation, as is the case with the statutory undertakers who have rights which allow them to lay mains in the highway, or it may be a common law entitlement, such as an occupier's right to the 'quiet enjoyment' of his property, or the right of every person to expect a contractor to take reasonable care when operating in a public place.

The ICE Conditions of Contract place the responsibility for dealing with most 'third parties' on the contractor. The programming and organisation of statutory undertakers' works, for instance, are commonly made part of the contractor's obligations through the inclusion of special conditions under clause 72, and clause 22 requires the contractor to indemnify the employer against claims for damage or injury 'to any person or property whatsoever'. Nevertheless, 'third parties' often deal with the resident engineer. Sometimes this is a deliberate decision (British Rail, for example, always communicate through the Engineer's organisation) and sometimes it results from understandable ignorance of the contractual arrangements (the motorist who has damaged his suspension on an unmarked ramp is unlikely to be interested in the niceties of clause 22). Where the conditions of contract provide for the contractor to handle particular 'third party' matters, the resident engineer must ensure that the proper procedures are followed, and refer all such issues to the agent. He should not, however, ignore these transactions entirely: in matters large and small, the employer's interests are best served by smooth, effective action, prompted by the resident engineer as necessary.

Members of the public are interested in civil engineering

projects for a variety of reasons. Some are near neighbours, who are disturbed by the construction noise and traffic, others may have their travelling arrangements disrupted. Amenity groups may be concerned about the effects of the scheme on the environment. Ratepayers and taxpayers have a very real interest in the public works they are financing. The resident engineer and his staff have a duty, to the employer and their professional colleagues, to ensure that good public relations are maintained throughout the contract. Although some information, such as the arrangements the employer has made with landowners for the purchase of their property or for access, is confidential, most reasonable enquiries should receive a prompt answer — in layman's language. When a scheme is likely to stimulate general public interest it is worth preparing an information sheet: a single page with a sketch plan, and a brief description of the works is sufficient, and can be produced cheaply for distribution to visitors, local libraries and any interested person.

For many people, however, the local press or radio will be the source of information on the contract. Too often it is only the problems which make news. Despite the widely-held view that journalists are interested only in scandal, disaster and crime, most local newspapers and radio stations rely for a large part of their output on the reports they receive of current events in the community. A story about a construction project, especially if there is an element of 'human interest' or the opportunity for a good photograph, will generally be welcomed and given reasonable coverage — provided the information is made available. Engineers who denigrate the press and point to instances of uninformed criticism, inaccuracy and remarks taken out of context are often unaware of how journalists work. Any story consists of facts, and, perhaps, opinion. If the reporter can obtain no authoritative statement and is denied any informed comment it is hardly surprising that he will present his editor with what he can get, even if it is a mixture of hearsay and his own views. If a problem arises on site, or if there is an accident, a garbled and unfavourable report is almost inevitable unless a proper press release is prepared. Engineers who will not talk to reporters 'because they always get it wrong' are simply playing their part in a vicious circle. Quotes are the lifeblood of journalism and anyone who speaks to the press must realise that, unless they have made some prior arrangement with the reporter, every word they say is on the record and may be used. However, journalists are usually scrupulous in honouring the convention that a statement which is 'on background' or 'off the record' will only be used to build up the text of the story and will not be printed as a quotation. The golden rule is to make this clear before giving the information—it is no good saying something sensational or highly confidential and then asking for it to be left out. The great benefit of such briefing is that the news item will include some basic facts and the reporter will know enough of the views of the site staff to produce a balanced story.

A contract was let to construct a new road through a residential area in a small town. The scheme had generated a lot of opposition because of its effect on the gardens of a terrace of houses and the arrival of the contractor was greeted with considerable apprehension.

One of the gardens affected by the works contained a greenhouse full of tomato plants. The agreed accommodation works provided for its replacement in a new position. The contractor took possession of the site, including this particular piece of garden, and set about clearance work, only to find that delivery of the new greenhouse would be delayed by several weeks. Well aware of the potential for bad publicity and a complete breakdown in relations with the local community, the resident engineer persuaded the agent of the advantages in 'working around' the old greenhouse. Excavation to formation level was carried out with the greenhouse, and its plants, standing on an 'island' in the middle of the works until its replacement was ready. Because the delay in delivery was the fault of the supplier, the contractor carried the cost of the disruption to his operations.

The local newspaper was contacted, a reporter was invited to the site and encouraged to take photographs. The result was a good-humoured piece, with complimentary remarks about the contractor from the tomato-grower and his neighbours, and some favourable general publicity for the project. Thus, by working *with* the press the original problem was turned to the advantage of the scheme. The alternative does not, of course, bear thinking about.

Chapter 9
On site: constructing the works

The purpose of a civil engineering contract is the construction of the works. Although the resident engineer and the agent, and all the members of their staffs, share this as a common aim they will not view the construction process from the same standpoint. The resident engineer seeks the best standards of workmanship and materials that can be obtained for the employer within the terms of the contract and for the agreed price. The agent strives to ensure that he fulfills the requirements of the contract in such a way that his costs are minimised. These differences in approach are always present and whatever the superficial similarities between much of what the resident engineer's team and the agent's staff do, their functions are nevertheless quite separate.

Superintendence and supervision

The agent superintends the construction of the works, the resident engineer is on site to supervise. At first glance, the terms may seem to be interchangeable, and indeed they are often confused, but they describe two distinct tasks.

Superintendence, as far as civil engineering construction is concerned, involves the active direction of the means by which the works are carried out. This covers organising resources, ordering materials, programming tasks and setting targets for both productivity and cost. Supervision, on the other hand, involves control over what will be accepted into the works. It includes assessing the suitability of methods, accepting materials, approving workmanship, determining the value of what has been done and the liability for any extra costs. This does not mean there is never any overlap. For instance, in choosing the best source of aggregate the agent will consider many of the factors which the resident engineer will later take into account when deciding whether or not to approve the source but their lists will not contain all the same items, nor will those items that are common be ranked in the same order of priority.

Superintendence is necessarily more active and positive than supervision for it is the contractor who has agreed to construct the works for the employer. The supervisor's role, however, should not be seen as predominantly passive and negative, for effective supervision requires more involvement than merely reacting to the actions and initiatives of the contractor. The process operates in four ways. The first of these is best described as 'troubleshooting' and is a continuous activity in which the resident engineer's team use their knowledge of the site and the design to anticipate problems and warn against them, watch out for potential mistakes and suggest means by which difficulties can be avoided. 'Troubleshooting' is perhaps the most positive facet of supervision and, when conducted successfully, the least conspicuous. It is not, however, an exhaustive check, not is it some kind of comprehensive insurance for the contractor, and it is largely carried on without formality. An indirect approach combining persuasion, advice and constructive criticism is usually very effective and can eliminate a great deal of direct, 'official' supervision, but it must never undermine, or interfere with, the contractual position, in particular the contractor's prerogative of deciding for himself how he is to execute the works and, in some cases, what mistakes he is going to make. Furthermore, the informal approach can only be sustained as long as matters can easily be rectified or reversed. When persuasion fails the resident engineer must respond to the situation and use his formal powers to disapprove, reject or instruct.

The contractor on a new by-pass was about to begin topsoil strip with scrapers and proposed to start in a meadow at one end of the site. A major land drain, to be replaced in the completed works by a system of new ditches, ran across the meadow. Although the drain was marked on the contract drawings, the agent had failed to appreciate its shallow depth which would inevitably result in its destruction by the scrapers, leaving the meadow without drainage and blocking the outfall from nearby farmland. The resident engineer pointed out the problem to the agent who quickly revised his detailed planning, sent the scrapers to another part of the site and brought a gang in to the meadow to excavate enough of the ditches to maintain the drainage of the area.

The second element of supervision involves overseeing or watching the execution of the work on the site to ensure that proper methods are used. In this activity the members of the site team, and the inspectors in particular, keep a constant check on the contractor's workmanship, sometimes using their own experience and knowledge of good construction practice as a yardstick but also applying the detailed method statements set out in the specification. Where clauses specify particular requirements precisely, the number of passes to be given to a layer of fill by a particular item of compaction plant, for instance, then supervision is a purely passive exercise — counting the number of times a roller traverses the surface of the fill. However, as the section on 'compliance' later in this chapter (p. 147) will indicate, not every requirement is so rigidly set down and not all of this kind of supervision is so simple. Although failure to meet the requirements of the specification means a presumption that the work is unacceptable, experienced site supervisors are aware that practical considerations can be just as valid as the details (or what the agent likes to call 'the fine print') of the contract documents. Satisfactory results may be achieved by using methods which are not exactly those described in the specification but which, when conscientiously applied and carefully supervised, represent a good alternative.

Most roadworks specifications contain provisions covering the rectification of pavement layers when the tolerances for level have been exceeded. In the case of lean concrete roadbase which has been laid too high, the usual requirements involve either re-trimming and re-compacting within two hours of placing or complete removal and reconstruction. When a pavement gang is working at a high rate of production the first option is often not available due to difficulties of access or the disruption caused by sending plant back from the laying area to do remedial work. When the finished layer is of sound construction and the size of the error and the area affected are within reasonable limits, site staff know that careful and limited breaking out of the surface layer can be successful. Such action would not be undertaken without full supervision and a careful check on the state of the roadbase layer left after treatment. The result can be more satisfactory than piecemeal excavation and reconstruction, and the method is employed for this reason and not because it 'lets the contractor off'.

Whether the operation is tightly specified or not, the presence of one of the supervisory team to watch its execution is always desirable. Unfortunately, the nature of site work is such that sufficient resources may not be available to give every operation full-time cover at busy periods and consequently the resident engineer may have to set priorities. A prerequisite of an effective system of supervision under these circumstances is information on what is happening on the site. The agent is under no general obligation to advise the resident engineer of the timing and location of every single operation he is undertaking, apart from the requirement (in clause 38 of the ICE Conditions of Contract) to give notice before any work is 'covered up or put out of view'. However, clauses are usually inserted in the contract documents to provide for the submission of a weekly programme and to ensure that certain activities of particular importance, such as concrete pours or compaction work outside normal hours, are subject to special arrangements for advance notice. By using these formal procedures and by relying on the informal contacts developed by his staff (the inspectors usually have excellent sources amongst the gangers and foremen), the resident engineer should be able to provide appropriate cover for every operation. The level of supervision varies with the activity: a major area of fill, a structural concrete pour or the working of a carriageway surfacing train require full time attendance and perhaps more than one person on duty; pipe-laying, trimming side slopes or the grit-blasting of steelwork can be covered by occasional visits and an inspection on completion, and one supervisor is able to look after several such operations. The effectiveness of the contractor's own superintendence is a critical influence on the resident engineer's strategy for supervision. Should the agent's staff be slack or insufficient in numbers, a closer watch will need to be kept on site operations if satisfactory standards are to be assured.

The third aspect of supervision is the checking of compliance with the standards of performance laid down in the specification. This may be done by the resident engineer's team acting on their own initiative, through the operation of the testing programme or by the agent formally submitting work for acceptance and approval. The supervisory staff follow up many activities closely and set up their own checks or carry out their assessment of the finished work as soon as it is complete: surface tolerances on carriageway construction, for instance, are normally measured as the work of laying the pavement proceeds; concrete finishes are inspected as the formwork is struck. Materials being brought onto the site and the end-products of various site operations are subject to testing conducted by the laboratory as a matter of routine: filter media is assessed for grading; concrete samples undergo fresh analysis to check the constituents and specimen cubes are made up and crushed to determine the compressive strength. Completed work due to go into the monthly valuation, or which is about to be covered up, may be put forward by the agent for the resident engineer's written approval: pipe runs are submitted for inspection of line, level and joints before backfilling commences; formwork and reinforcement is presented for final approval before concrete is placed.

The essence of this form of supervision is speed and accuracy. Clearly these qualities are vital when work is to be covered up or forms the foundation for the next phase of construction, but they are also necessary when the check is on a continuing process. There is no value in detecting errors or deficiencies too late to have them rectified nor in indicating corrective action on the basis of inaccurate data. Success depends heavily on the efficiency of the site laboratory and the competence of the resident engineer's team in the basic skills of measurement and levelling. However, these are not the only ingredients, for on a busy contract the rigorous but time-consuming procedures of testing and physical checking must be supplemented by faster but more empirical methods based on judgement and experience. These range from the visual inspection of the concrete in the chute of a truck-mixer, which is all a long-serving inspector needs to tell him how much water can be added without exceeding the required slump, to the jabbing of the heel into the formation (the famous 'boot test') which is so often used on road contracts to estimate the strength of the sub-grade when there is no time to measure the *California Bearing Ratio* (CBR) value.

The fourth part of the supervisory process involves approving and monitoring the contractor's proposals. Although it is discussed last here, it is generally initiated in the early stages of the project when the agent formally submits details of his methods of working, sources of materials, and suppliers of components. In some cases trials are specified, for instance, concrete finishes are usually demonstrated on sample panels, and in any event the resident engineer does not simply consider the proposals 'on paper', but will arrange to inspect any off-site locations such as quarries or fabricating yards and to receive examples of items such as guardrails. When similar work or materials have been incorporated in other schemes, it is useful to examine the finished product and seek any practical advice which may be available. The approval, which may be qualified, becomes the yardstick against which actual performance is compared as the work proceeds.

Throughout the progress of the contract, therefore, the resident engineer gives his approval, expressly or by implication, to the contractor's materials, methods and workmanship and to any proposals the agent may submit regarding the execution of the works. What comfort can the agent take from the knowledge that the resident engineer and his team are exercising their duties of supervision? The answer, unfortunately for the agent, is: none at all. Most agents and their staffs would hold the view that, once the resident engineer has given his approval, and provided no deliberate deception is involved, his judgement should stand and if it is modified, withdrawn or overruled then the consequences should be borne, or at least shared, by the employer. The resident engineer, however, cannot operate such a principle and will require the liability for any extra cost in such circumstances to be carried by the contractor. This attitude is usually greeted with incredulity and dismay, sometimes genuinely expressed, for many contractor's personnel believe every decision of the resident engineer represents a firm committment by the employer. Similarly, members of the resident engineer's team are often unsure of their position when faced with a case in which they are personally involved and for which they feel a degree of personal responsibility. Particular cases, however much uncertainty they may arouse when viewed in a limited context, cannot detract from the general contractual principle that neither the Engineer nor the resident engineer owe any duty to the contractor to prevent, or even to warn against, any mistakes he may be making.

This hard rule stands on two foundations: firstly, the legal status of the Engineer (or the resident engineer acting under delegated powers) and the contractor; secondly the particular terms of the ICE Conditions of Contract. There is no contract between the Engineer and the contractor, and thus there can be no direct obligation or promise on which the contractor can rely. The Engineer does have a duty 'to act fairly as between the parties', but this refers to his obligation to remain unbiased when making decisions on contractual matters. It does not mean that in his supervisory capacity every act of approval positively discharges the contractor from further liability in the particular matter, nor that his silence implies approval. Of course, no responsible Engineer would deliberately withold any expression of concern, doubt or disapproval, nor expect his resident engineer or any other member of the site team to do so, for that would undoubtedly be 'unfair' in ethical, if not legal, terms. But neither can the Engineer be considered as the contractor's specialist adviser, offering expert opinion on which the contractor would place particular reliance. The contractor has already, in tendering for the project, affirmed that he is experienced and skilled at the work he is undertaking. He cannot argue with conviction that he relies upon the Engineer's judgement to guide him in the execution of his business and thus enjoys the 'special relationship' which the law requires to establish indirect liability. The ICE Conditions of Contract reinforce these general principles with two sets of specific provisions. The following clauses state the obligations of superintendence.

Clause 8	The contractor is to provide everything needed for the construction, completion and maintenance of the works and 'shall take full responsibility for the adequacy, stability and safety of all site operations and methods of construction'.
Clause 15	The contractor must provide 'all necessary superintendence' which must be given by 'sufficient persons having adequate knowledge of the operations to be carried out ...'.
Clause 16	All the contractor's staff must be 'careful, skilled and experienced in their several trades and callings'.
Clause 17	The responsibility for the 'true and proper setting-out of the Works' rests entirely with the contractor.
Clause 20	The contractor 'shall take full responsibility for the care of the Works'.
Clause36	'All materials and workmanship shall be of the respective kinds described in the Contract'.

The following clauses limit the duty of supervision.

Clause 14	Approval of the contractor's progamme and methods, 'shall not relieve the Contractor of any of his duties or responsibilities under the Contract'.
Clause 17	'the checking of any setting-out or of any line or level ...shall not in any way relieve the Contractor of his responsibility for the correctness thereof'.
Clause 39	Any materials or workmanship which is not in accordance with the contract can be ordered to be removed or re-executed

'notwithstanding any previous test thereof or interim payment therefor' and 'failure of the Engineer or any person acting under him ...to disapprove of any work or materials shall not prejudice the power of the Engineer or any of them subsequently to disapprove such work or materials'.

Clause 60 — The payment in a certificate for work done or materials supplied does not imply full satisfaction, for any such sum may be subsequently deleted, corrected or modified.

Of all these terms, clause 39 is the most significant. It is generally held to apply not only to cases of the resident engineer's 'failure to disapprove', but also to instances where approval has actually been given.

Taking all these provisions together it is clear that the resident engineer is entitled to the presumption that the contractor is aware of his obligations and knows what he is doing. If at any stage that presumption is proved ill-founded then the resident engineer must take action, and the contractor must take the consequences. Whatever 'errors' can be attributed to the resident engineer, for example, absence of a supervisor during an operation, failure to detect, notify or disapprove of any unsatisfactory or defective work, inadvertent approval or acceptance of work later found to be not in accordance with the contract, the responsibility rests with the contractor unless expressly excluded by the ICE Conditions of Contract.

Compliance

The job of supervision requires the resident engineer and his team to make a series of judgements about the contractor's materials, methods and workmanship. In every case, the resident engineer has to assess whether the contractor has complied with the requirements of the specification, or, if there is no detailed statement of method or performance, with the accepted standards of good practice.

The means of assessing compliance depend upon the way in which these requirements have been expressed.

(1) Where the specification is strict and mandatory, such as when an upper limit is applied to the moisture content of imported clay fill, the process is automatic and the outcome is directly determined by a straight comparison against the standard.
(2) When 'deemed to satisfy' provisions are used, for instance by listing the names of acceptable component suppliers, compliance is effectively pre-determined and no individual assessments are necessary.
(3) Certain tolerances may be permitted so that, for example, the actual dimensions of a pre-cast unit may vary from those shown on the contract drawings by a specified amount, thus establishing a range of compliance against which individual results are assessed.
(4) The standards of compliance may be statistically based when large quantities of materials, for example structural concrete, are to be assessed. The process can take into account the consistency of the end product instead of focusing narrowly on single samples and can also identify trends so that adjustments can be made in advance to maintain satisfactory standards.
(5) The requirement may be expressed as an exclusion or prohibition but subject to a discretionary power of acceptance, as when material above a stated moisture content is declared unsuitable for use in earthworks 'unless otherwise permitted by the Engineer': this allows the initial assessment to be followed by a 'second opinion' based on the supervisor's own judgement.
(6) The specification may state that work is to be done 'as required by the Engineer' or 'to the satisfaction of the Engineer', as, for instance, with the cleaning and making good of re-useable formwork, or there may be no detailed provision set out at all. Compliance then becomes a matter for the supervisor's discretion, subject, however, to the customary standards and practice of the construction industry.

The extent of the discretion which supervisors are allowed to exercise increases from nil to almost complete. Compliance via a process of automatic approval or rejection lies at the bottom end of this range and no room for manoeuvre is intended. Nevertheless, the agent may seek to obtain the resident engineer's agreement to some relaxation, perhaps combined with an offer of a reduced rate for that item, and support his proposal with convincing arguments about progress or value for money. Although it would be naive to suggest that offers of this kind are always rejected, or indeed that they are always suspect, for many are made in good faith, it is only in exceptional circumstances that a departure from the strict requirements of the specification can be justified. To waive the standard of compliance when it has been so precisely stated is, in effect, to tamper with the contract and the ICE Conditions of Contract give no such authority to the supervisory staff, indeed, clause 13 specificially requires the contractor to work 'in strict accordance with the Contract'. Nor should the device of allowing a departure from the specification to go ahead 'at the Contractor's risk' ever be used, for that abdicates proper responsibility for supervision.

The use of 'deemed to satisfy' provisions in the specification gives the contractor a clear indication in advance of what materials or methods will be accepted. The agent has only to ensure that his aggregate comes from one of the specified sources or that his compaction plant is chosen from the permitted list to know that it will receive the resident engineer's approval. This theory is subject to practical modification. Automatic approval depends upon a presumption of 'fitness for purpose', so if the aggregate is delivered contaminated with clay or the compaction plant is not in proper working order then they will be rejected in spite of their superficial compliance with the specification. The likelihood that an experienced construction firm will be able to offer alternatives outside those listed in the specification is recognised in the wording of many 'deemed to satisfy' provisions which conclude with the phrase 'or similar approved'. The specified sources or methods nevertheless provide the yardstick against which both the agent and the resident engineer can judge the suitability of such alternatives.

The application of tolerances to the standards of compliance invariably produce occasions when the results fall just outside the range of acceptance. The agent will naturally press for approval and in seeking to persuade he will, of course, stress how small is the departure from the tolerance. If, for example, a measurement can be within plus or minus 10 mm of the dimension on the contract drawings and is 12 mm short, this will be described as 'only 2 mm out'. What must be remembered is that the tolerance expresses the maximum permitted range of error within which approval can still be given – in the example above the dimension is not 2 mm but 12 mm 'out'. The nearer the result lies to the end of the tolerance range, the less satisfactory is the product and the limit must be taken, subject to occasional exceptions, as the point at which satisfaction ceases. There are no tolerances on tolerances.

Similar disputes can arise over averages or more sophisticated statistical expressions of compliance. The purpose of these techniques is to minimise the effect of extreme or 'rogue' results by including them in a large population which is intended to provide an overall view of the quality of the product. These results are nevertheless significant and cannot be characterised as 'unrepresentative' for they give a valid indication of the variability of the specified parameter. When a set of results fails it is often possible to identify some conspicuously bad figures which may be seen, not necessarily correctly, as the only reason for the failure. The agent may point out how easily the test results would satisfy the specified requirements if only these few 'wrong' figures were eliminated; after all, he may argue, they clearly represent errors in sampling, testing or measurement and should not be taken into account. The resident engineer has to consider whether these are truly errors or valid results which should, indeed must, go into the calculation if the test of compliance is to be fairly applied.

When clear opportunities for discretion are offered, the resident engineer must take care to be reasonable, both to contractor and employer. In such cases he must be prepared to consider suggestions which are different in detail or even in character, provided there is no extra cost (and this includes not only initial cost but also running costs), no general loss in performance or reliability and no detrimental effect on appearance where this is a relevant consideration. Conversely, where something is prohibited by the specification unless accepted by permission of the resident engineer, it is similarly reasonable to assume that under normal conditions the prohibition will be confirmed, and only in special circumstances will a relaxation be permitted.

The widest discretion in assessing compliance is given when the specification requires the contractor to 'satisfy' the resident engineer that the workmanship or materials is of an acceptable standard. Even when the specification is silent, clause 13 of the ICE Conditions of Contract, which requires that all work is 'to the satisfaction of the Engineer', still applies. In all such instances the resident engineer cannot unreasonably withhold his approval, nor can he expect the agent to accept the extra cost of carrying out work in a manner which no experienced contractor could have foreseen. Resident engineers do not have unlimited discretion, therefore, and must not introduce unusual or unnecessary stipulations unless prepared to recommend that the employer pays the bill.

Whenever the site staff have to deal with discretionary compliance, they are likely to hear a great deal about what their colleagues on other contracts have been willing to approve. 'It's always been accepted before' and 'I've never been asked for that until now' are two well-used phrases in every agent's vocabulary, and however persuasively they may be delivered it is important to remember that the way in which resident engineers and their teams on other sites exercise their discretion is not necessarily of any contractual significance. Although it may be evidence of a standard practice so widely used that it amounts to an implied term, and is thus virtually binding, it is usually nothing more substantial than a useful pointer to what may be considered as reasonable. Supervisors should not allow their judgement to be swayed by the agent's helpful reports of what other, allegedly more experienced, site staff have accepted, or are said to have accepted, elsewhere. If there is doubt, it is of course only prudent to seek advice, but not from the agent.

Approval and acceptance

In most cases the specification does not lay down an absolute standard of compliance but offers a range of acceptability, allowing the contractor a choice of how to satisfy the requirement. However, whilst much of the contractor's planning can be seen as a series of choices taken within the framework of the specification, the agent's approach may not always follow such a process exactly. For example, no conscious decision may be made other than to try and stay within the limits of the specification, or the contractor may find his choice is predetermined by what is available locally. On the other hand, there may be a deliberate choice as when a source of cheap but 'marginal' material is adopted for economic reasons or an expensive but high-quality supply is selected for its benefits in expediting progress. In every case, whether the process goes by default, is influenced by external constraints, or consists of a conscious decision, the result is nevertheless a choice in the sense that other options are excluded. Furthermore, by whatever process the choice is made, it is rarely a self-contained decision. Consequential effects are bound to follow and the earlier the point of choice in any process, the more far-reaching and pervasive the implications.

The mechanism of choice can be depicted as a cone (see Figure 8). The broad base represents the starting point at which a wide range of choices are available, the elimination of options and the pressure of external factors progressively reduces the area of choice as the process or project continues until the final result, represented by the apex, is attained. The object is to reach a result which is the same as the specified target, and thus acceptable, but incorrect choices, the action of external restrictions or pressures and the passage of time may distort the cone. Thus in order to achieve the target the agent must make a series of choices which keep him within the specified cone. The closer he comes to its outer surface, the greater the risk that the next choice or some external agent, such as the weather, may put him outside the cone altogether.

Clearly the Engineer has an interest in whether or not the contractor produces the correct result, but to what extent should he involve himself in the earlier stages of the process?

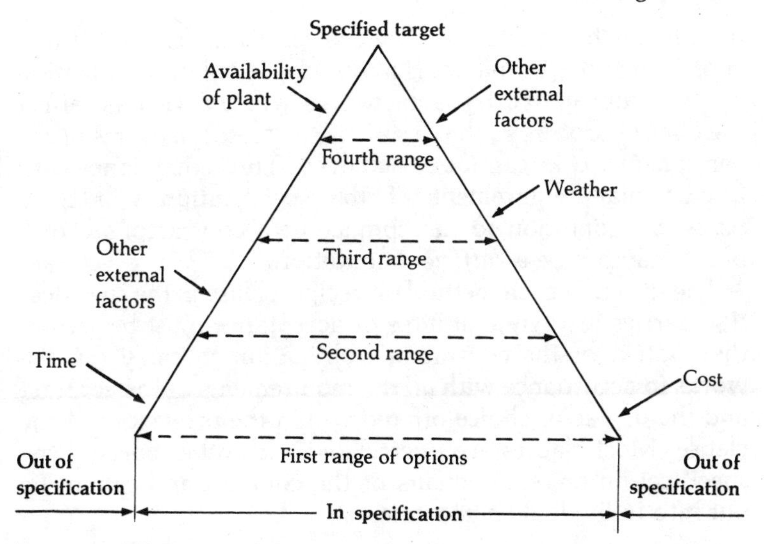

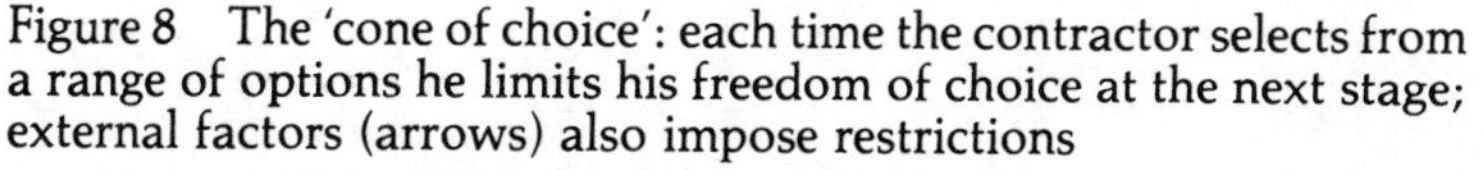

Figure 8 The 'cone of choice': each time the contractor selects from a range of options he limits his freedom of choice at the next stage; external factors (arrows) also impose restrictions

There are always two aspects of the Engineer's supervisory role: one is limited in scope and consists of comparing the contractor's efforts with the requirements of the specification; the other is the duty to ensure that the permanent works are satisfactorily constructed. The two aspects are interlocking and it is unwise to believe that if the first is given careful attention then the second will look after itself.

The Engineer must have due regard for the various tolerances, permitted variations and ranges which the specification encompasses. On occasions, he may find himself unable to agree with (or even to understand) the contractor's choice, or consider that the contractor's chosen technique does not produce the optimum result, but preference, even when backed with engineering judgement, is not grounds for rejection. If the choice lies within the limits of the specification, there is a presumption that it is acceptable. However, that presumption is not absolute, and here the second aspect of supervision, ensuring the adequacy of the permanent works, comes to the fore, because the test of compliance cannot always be conclusive. The special characteristics of any chosen methods or materials may, when used in combination or when subject to external factors such as the weather, affect the overall process so that a satisfactory result in terms of the permanent works cannot be achieved. Thus compliance with a particular requirement of the specification will be a *necessary* condition for acceptance of a contractor's choice but it may not be a *sufficient* condition.

The idea that each method or 'recipe' clause in the specification carries its own guarantee of acceptance must be seen in the context of the contractor's obligation to carry out the works in accordance with *all* the requirements of the contract and the degree of choice offered within the provisions of the clause. Most clauses are interrelated with other parts of the specification or requirements of the contract and cannot be put into individual compartments.

Example (A): surface texture on hot rolled asphalt

Hot rolled asphalt wearing course supplied and laid to the Department of Transport specification is governed by a number of clauses and an associated British Standard which set out to control the constituents, the stiffness of the mix, laying technique and the regularity and texture of the finished surface. The final requirement is for a wearing course of acceptable material laid so as to provide a smooth-riding and durable surface with a minimum standard of texture depth. In order to achieve this the contractor must work through a set of choices and external factors which can be sub-divided into three groups:

(1) characteristics of the asphalt;
(2) characteristics of the chippings and their application;
(3) operational constraints.

(i) Characteristics of the asphalt. The asphalt acts as a matrix in which the basic elements of the texture, the chippings, are set and held. Its performance depends on the properties and proportions of its constituents: binder, filler and aggregates.

By selecting different types and varying the quantities, a wide variety of asphalts can be produced ranging from very soft to very stiff.

The softer the material the morc firmly the chips are embedded and the better the compaction under the roller. The disbenefits come from the mobility of the asphalt during rolling, which produces a tendency towards an irregular riding surface and excessive chip penetration, the latter resulting in poor texture depth. The advantages of a stiff mix are its durability and resistance to deep penetration of the chips, which combine to give good surface texture; the disadvantages lie in the risks of inadequate chip embedment, leading to early loss of texture, and of under-compaction which can lead to complete failure of the wearing course.

(ii) Characteristics of the chippings and their application. The specification regulates the quality of the stone: its grading, size, shape and resistance to polishing and abrasion. The rate of spread and distribution are not so strictly controlled but are nevertheless of great importance.

A contractor working with a relatively soft asphalt may find that he can only achieve a satisfactory surface texture by chipping at a rate in excess of that recommended in the British Standard, although the asphalt conforms with the requirements of the specification. Chipping at this 'enhanced rate' makes up for the high proportion of stones 'lost' into the asphalt layer, but it must not be so high as to cause excessive bunching, crushing or inadequate embedment. The contractor is not without other options: he could modify the proportions or change some of the constituents of the asphalt and, whilst still complying with the specification, produce a stiffer mix.

(iii) Operational factors. The organisation and control of the laying operation are very important. A consistent supply, in terms of mix proportions, delivery temperature and rate, will significantly improve performance. Similarly, if the previous surfacing layer has been laid to a high standard of regularity, the wearing course will be of uniform thickness and good chip embedment and compaction can be achieved without the problems of constantly adjusting the amount of rolling. Sufficient rollers must be available at all times, and the supply of

chips to the chipper, and the action of the chipper itself, must be reliable and regular. The agent who cuts corners in this part of the operation is choosing, consciously or unconsciously, to risk rejection of the finished product. The weather must also be taken into account. Cold or windy conditions means a stiffer layer of asphalt; warm sun produces a soft and mobile material. The asphalt, the chipping rate and the laying routine must be varied accordingly.

The cone of choices and constraints is shown in Figure 9. It can be seen that the direct path A – Z is the safest route, since the agent's decision to keep to middle-of-the-range options gives him ample room for manoeuvre when external factors work against him. The path represented by the line B – Z (asphalt of low stiffness/high delivery temperatures/average-to-low-chipping rate/average compaction) is a risky one, for external factors, such as a run of poor quality chippings or a spell of hot weather, could easily put the target out of reach. The resident engineer will have to examine the agent's choices

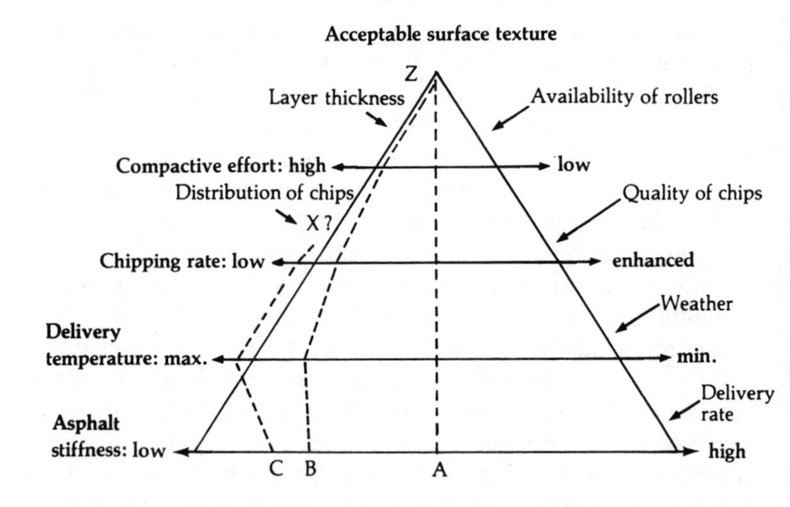

Figure 9 'Cone of choice' for surface texture on hot rolled asphalt: the four specified parameters for surfacing (bold type) successively limit the contractor's freedom of choice; he must also allow for the effect of external factors (arrows) as he works along his chosen route (broken lines) to the target

and actions at each stage to assess their likely effect on the permanent works. The incomplete path marked as C − X (low stiffness/maximum delivery temperatures/minimum chipping rate) shows how the agent's early choices can leave him with no option but to use very light compaction to bring himself back inside the 'cones' thus running a significant risk of under-compaction of the asphalt mat and subsequent failure of the surfacing in use.

Example (B): free-draining material

The normal specification for free-draining material is a 'recipe' consisting of a 'deemed to satisfy' list of materials, and a wide range of permitted grading. This grading envelope allows a significant proportion of fines which, in the case of relatively smooth bed gravels and similar sources, can include enough clay or silty clay to cause problems in wet conditions when the behaviour of the fine fraction tends to govern the performance of the material as a whole. Pore water pressures can develop to the extent where compaction of subsequent layers, and sometimes even the passage of plant, is impossible and the material must either be left to drain, or assisted in drying-out by the provision of grips or layers of clean stone.

The contractor has complied with the specification, however, and may argue that his obligations have been fulfilled. Surely any delay or extra work are now the responsibility of the employer, whose specification is so clearly at fault? The answer must be 'no', for it is not the specification which has caused the problem but the contractor's own choices, his programming decisions and the weather, which, unless exceptionally severe, is a risk he must bear.

The specification offers the contractor the opportunity to select from a wide range of materials, all of them capable of performing adequately in the permanent works, and he must be presumed to have chosen with deliberation, taking into account the consequences. Suppose for instance, that a smooth, rounded gravel containing a relatively high proportion of clay is obtainable locally: it complies with the specification, is available at short notice and is cheap; it is also weather-susceptible and provides little in the way of mechanical interlock. At a much greater distance there is a source of crushed rock: it, too, complies with the specification, but must be ordered in advance and will have to be stockpiled, is more expensive and involves organising long-distance transport; however, this material provides a high degree of stability and will drain rapidly in all weathers. In choosing the first source, the contractor must be considered to have weighed these and any other relevant factors and decided that the cost of any likely delays or extra works attributable to the nature of his selected material are worthwhile in the light of its availability and basic cheapness.

The 'cone of choice' for this example (see Figure 10) is of a different type from that for surface texture, not because of the

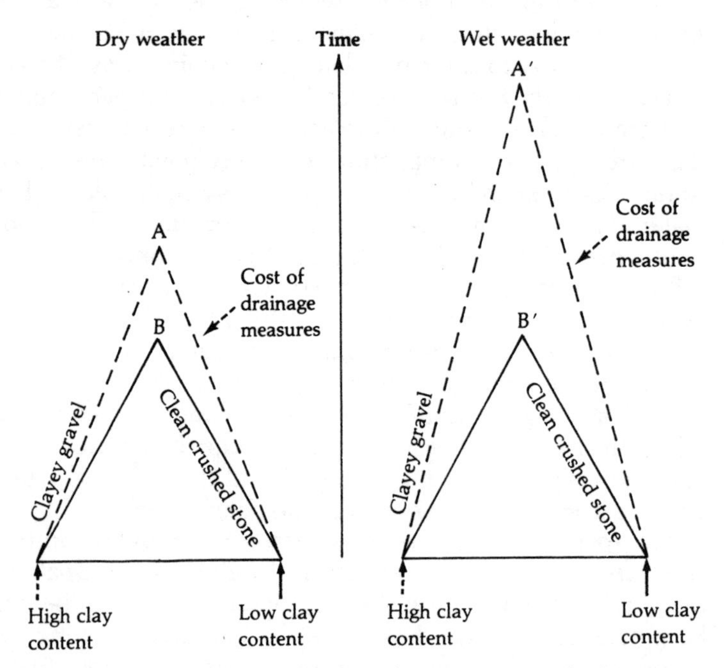

Figure 10 'Cones of choice' for free draining material: the targets for dry conditions (A,B) and wet (A′,B′) are the same — an acceptable completed fill; the choice of a high clay content (broken line) always involves more time, which can only be reduced by spending money on drainage grips etc.

numbers of options or external factors involved but because time is of such major importance. Although the target for acceptance is always the same, a satisfactory base to support the next layer of construction, the different cones represent the different times required to achieve that target. In both dry and wet weather the cones for the crushed stone are of the same height, as the time required for laying and proceeding to the next stage is not weather-dependent. This is not so with the clayey gravel, and the additional heights A – B and A′ – B′ represent the extra time which varies with the weather and the amount of extra works (drainage grips etc.) which the contractor intends to provide.

Workmanship

The specification will not contain detailed descriptions of such techniques as bricklaying, carpentry, welding or the hand-laying of mastic asphalt. Each of these, and the many other specialised operations on a construction site, is more a craft than an industrial process and requires skilled and experienced personnel for effective supervision. To maintain good standards of workmanship a knowledge of the 'tricks of the trade' is essential. This does not mean that only an ex-bricklayer can supervise brickwork (although the preferred choice would be a former 'brickie'), but it does require the supervisor to have a practical grasp of the skills involved in the work, to know what can go wrong and how it can be put right.

The inspectors have a leading role to play in this aspect of site supervision. Often former skilled tradesmen themselves, but in any event experienced in the practice rather than the theory of civil engineering, they are able to guide, advise, warn, persuade and cajole the contractor's men into producing the required standard. Much of their supervision is done informally, through the close contacts they develop with the site labour force, the gangers and the foremen and depends on respect for the judgement and fairness of individual inspectors. The relationship is a delicate one and, as the assessment of workmanship is partly subjective, it should not be put at risk unless an inspector is clearly wrong.

Inspectors are not infallible, however, and sometimes their natural inclination towards the 'muddy boots' side of engineering goes with a rather sketchy knowledge of the more recent additions or modifications to the specification. Their standards, always high, are occasionally somewhat higher than those the employer had in mind when drawing up the contract documents, and there are times when an inspector's requirements amount to unforeseen additional work and therefore extra costs. The resident engineer must treat his inspectors, and their reputations on the site, with care, but he must always be prepared to have the courage of his own convictions.

Setting-out

Establishing the accurate position, line and level of all the various parts of the works is a vital part of the construction process, and many specialised textbooks are available to cover this area of expertise. As far as the task of supervision is concerned, however, there is only one simple rule to be learned, which is stated in unequivocal terms in clause 17 of the ICE Conditions of Contract: as long as he is provided with current and sufficient data in the contract documents, any errors in setting-out and their consequences are entirely the responsibility of the contractor, no matter how caused or when detected.

It is, of course, a wise precaution to examine any calculations the contractor's staff may do. The specification often includes a requirement that the contractor establishes master survey stations on the site and checks them against the survey control used by the designer before detailed setting-out can begin. Similarly, all physical setting-out such as level pins, traverse stations and pegged alignments ought to be checked before it is put to use. This ensures against any incorrect data being used by the contractor and also serves as an early warning of any difficulties likely to arise because of a lack of information.

The double-checking process helps to pick up any errors of mathematics or surveying before work is started, but in contractual terms it has no status and does not relieve the contractor of any of his obligations. Even minor errors in setting-out or taking dimensions off the contract drawings can have serious repercussions. Approval of any calculations, alignments or levels, even if accompanied by a signature on a form or given some other kind of 'official' recognition, does not remove the contractor's liability. Failure to detect mistakes in this very important activity can be an embarrassment and reflects badly on the resident engineer's team if it happens frequently, but, as in other matters of supervision, there is no duty to protect the contractor from the consequences of his own errors.

Testing

On a large contract the laboratory is a very busy place indeed and the resident engineer must ensure that its resources are properly employed to provide the information which is essential to the control of the works. In performing this task the resident engineer must strike a difficult balance between the legitimate role of the laboratory in seeking out and confirming the existence of any improper work and materials and the practice of directing the testing programme towards what may be called 'contractor-bashing'. The latter course can only compromise the whole output of the laboratory and undermine its potential for providing a mutually acceptable source of quality control.

In general terms testing on any contract may be divided into three categories:

(1) *Compliance testing* is carried out to ensure that materials satisfy the requirements of the specification. Typical examples are cube strengths for concrete and moisture contents for earthworks materials. In the initial stages of a contract compliance testing is carried out frequently to establish a satisfactory basis for evaluating and accepting materials, particularly those coming from sources outside the works or the main contractor's organisation. At a later stage the frequency of testing can be reduced when the supervisory

staff are generally satisfied as to the acceptability of the material concerned. At that stage compliance testing becomes control testing (see below) although exceptions will continue to occur when staff consider that material is clearly outside the specification or likely to prove troublesome.

(2) *Control testing* is carried out to ensure that satisfactory standards which have already been set or otherwise established are maintained. A typical example can be found in surface texture testing which is conducted initially at a high frequency to establish the method of working which will produce satisfactory results and thereafter much less often as staff become able to accept work on the basis of their own judgement.

(3) *Record testing* is carried out to establish a comprehensive review of the quality and/or performance of all the material provided or used in executing the works. Random sampling and testing is conducted continously regardless of whether the materials are judged to be suspect or not and thus many samples are taken from materials which are known to comply with the specification. The results are intended as a record but may be used for control when they reveal problems.

On a normal contract the laboratory's efforts are directed almost entirely at compliance and control testing. It is unusual for a comprehensive system of record testing to be set up as this would be beyond the resources of a typical site laboratory. Consequently, sampling is necessarily biased towards materials which are considered to be suspect, and throughout the duration of the contract supervisory staff accept materials, both imported and won on the site, into the works on the basis of a visual assessment or some other empirical and unrecorded test. Laboratory results, therefore, are not representative of the whole range of quality found in the materials on a contract. This approach is followed in respect of most major site activities, particularly earthworks, and allows the work to proceed at a reasonable pace without unnecessary delays. The expense and effort of full-scale record testing is justified only when the unusual or uncertain nature of the works makes a claim likely, or when it is known that the contractor is submitting a claim to which the results of a comprehensive sampling and testing programme would be relevant. Claims in respect of ground conditions fall into this category and form the largest single class of substantial contractual disputes. The value of a complete and representative record in settling such a case is clear and explains the emphasis in the ICE Conditions of Contract on the proper notification of claims.

During the progress of the works the agent should be encouraged to associate himself with the laboratory's work. He should be invited to nominate a member of his staff to act as a contact with the materials engineer, and offered the opportunity to witness any sampling or testing in which he may be interested. In this way, the agent can judge for himself the efficiency of the staff and be satisfied that their results are accurate and impartial. There could, under these circumstances, be no objection to providing duplicate copies of all laboratory results to the agent, who would in return be expected to present a prompt and reasoned criticism of any which he considered inaccurate or unrepresentative. This arrangement means that the laboratory does not sample and test in secret and that the agent can obtain, on request, full disclosure of results.

The essential pre-requisite of such co-operation is the agent's involvement. If, however, the agent appears to be unwilling to associate himself with the testing programme, the resident engineer is entitled to be cautious, and to consider the possibility that the contractor might be keeping at a distance deliberately, to retain the freedom to criticise the laboratory or reject its records in the event of a dispute. Without involvement, the agent cannot expect full disclosure and the resident engineer need release only those results which are necessary to support warning or rejections.

Care of the works

During the construction period and, under the ICE Conditions of Contract, for 14 days after the issue of the certificate of completion, the contractor is required to assume 'full responsibility for the care of the works'. Any damage, loss or injury 'from any cause whatsoever' must be made good or repaired by the contractor at his own cost in such a way as to ensure that the permanent works are completed fully in accordance with the contract.

The only exceptions permitted by the ICE Conditions of Contract are in respect of certain clearly defined circumstances known as the 'Excepted Risks'. Clause 20(3) provides the details which may be summarised as follows:

(1) war or other acts of external force;
(2) riot, rebellion or other acts of internal force;
(3) nuclear accidents;
(4) supersonic shock waves;
(5) a cause due to the occupation by the employer (or anyone working for him) of any part of the permanent works;
(6) a design fault.

All these exceptions are normally required by the insurance market and are common to most 'All Risks' policies. As the contractor cannot insure against them, they must be borne by the employer. The last two are the most important. The use, however temporary or minor, of the works by the employer absolves the contractor from liability for any damage arising and so the employer should consider his position (and check his insurance) before occupying or using any part of the site. It may be impossible to avoid such a risk, as for instance when the employer has to bring in another contractor to do separate work on the site such as diverting services, and in such circumstances every effort should be made to minimise the extent and duration of the occupation. The resident engineer and his staff may be in a difficult position as they may have only limited control over the employer's activities in this situation; nevertheless, by assuming, as far as possible, an advisory and co-ordinating role the site staff can usually prevent any unnecessary difficulties.

Design faults are sometimes clearly identifiable and when

damage or loss arises from such a cause any dispute is usually over the cost and not the liability. Not all cases are so simple, however, and the burden of proof is on the contractor to show that the damage arose wholly or principally from a design fault or one of the other 'Excepted Risks'. This may appear to set the contractor a difficult task, but it must be remembered that it is only necessary for him to show that the main cause, and not the sole cause, of the damage was a design fault to recover all his losses. Furthermore, it is accepted that proof can be established by a process of elimination: thus if it can be shown that, for instance, particular damage could not have been caused by actions or events within the contractor's responsibility, then the cause is presumed to be a design fault and thus an 'Excepted Risk'.

Although this interpretation of the standard of proof required to establish the presence of a design fault makes this 'Excepted Risk' very wide in its application, the contractor's responsibilities remain extensive. Thus the need to keep a careful check on the contractor's operations, and to maintain comprehensive records, is essential to the proper resolution of disputes over liability for the care of the works.

The action of the weather is a contractor's liability, as is vandalism and any other accidental damage caused by adjacent landowners, occupiers or members of the public. The phrase 'any cause whatsoever' covers almost every eventuality, even damage caused by the employer's own negligence, provided it is not associated with any use or occupation of the works.

A contractor was building a jetty. During the construction period the employer did not make any use of the works but one of his vessels, passing in the vicinity and handled negligently by her master, collided with the jetty and caused substantial damage. The contractor was liable for the cost of all the repairs which were held (the matter was finally settled by the courts) to be covered by his obligation to take full responsibility for the care of the works. (This is the case of Farr *v* Admiralty (1953) 2 All E.R. 512.)

The major cause of damage, apart from the weather, is the contractor's own activities. Haul routes for materials which run immediately adjacent to, or over, parts of the works are a frequent source of problems. The operations of sub-contractors, who sometimes press on with their own work regardless of, and even in spite of, the damage it may be doing to other sections of the job, are another common difficulty. The potential on a construction site for damaging one part of the works whilst executing the next is enormous. The use of compaction plant provides many typical examples, ranging from the puncturing of waterproof membranes by rammers working on bridge backfill to the displacement of kerbs by deadweight rollers working on carriageway surfacing. Most agents recognise these and other problems, work hard to prevent them and accept responsibility for the consequences when damage does occur. Nevertheless, the site team must be alert to detect any instances of actual or potential damage and ensure that they are properly documented and brought to the agent's attention. It is particularly important that any action, or inaction, likely to cause difficulties later in the construction period is noted at the time, so that a clear cause-and-effect link can be drawn. Careful judgement is needed in matters of this kind as if the site team 'cry wolf' too often they will lose credibility with the agent and his staff.

The contractor on a new dual carriageway had completed a section of embankment and installed the drainage at the top of the batter slopes. The embankment was then used as a haul road by laden motorscrapers carrying filling material to other parts of the works. The resident engineer formally drew the agent's attention to the way in which scrapers were being allowed to run very close to the line of the drain, pointing out that this failure to take proper care of the works was likely to result in damage both to the pipes themselves and also to the adjacent fill due to leakage of water.

Damage to the drain was identified and the contractor agreed to repair it. Soon afterwards a substantial slip occurred in the embankment alongside the damaged section, the fill being in a saturated and unstable condition. The contractor sought to recover the cost of the repairs to the embankment and also to use the failure as particular evidence in support of a general earthworks claim which he was pursuing. The resident engineer was able to point to the record and reject the contractor's arguments on both submissions.

Chapter 10
Variations

Few construction contracts run their course without the need to modify the work shown in the contract documents. The discovery of ground conditions in one part of the site different from those originally anticipated, the non-availability of a particular specified component and the introduction of extra working load requirements on a structure are typical examples of the events which initiate variations to the contract. Like so much of modern civil engineering, variations are not what they were. Few resident engineers would contemplate following in the footsteps of Sir John Hawkshaw, who took over as engineer-in-charge of the Severn Tunnel in 1879 and issued, as one of his first acts, an order to construct the section of tunnel under the river at a level 15 feet below that shown on the drawings!

Contractors expect variations, but rarely welcome them, and it is inevitable that some friction will be generated by certain changes which are, by their nature, extent or timing, less welcome than others. Nevertheless, by thinking ahead, by a proper appreciation of what the contract allows and what consequences can follow and through the correct use of procedures, the site staff can at least ensure that the problems of ordering and valuing variations are minimised.

What constitutes a variation?

The Engineer (or his representative) is authorised by clause 51 of the ICE Conditions of Contract to order variations to any part of the works. Employers can write restrictions into their agreement with the Engineer limiting the extent of any variation (the standard Association of Consulting Engineers form contains a general provision) and the resident engineer may also have similar limits imposed on him by head office. Such safeguards are usually applied loosely and the need for supervisory staff to act on their own initiative in urgent cases is invariably recognised, as is their independent duty to do what is necessary to secure the successful completion and operation of the works.

Clause 51 is widely drafted and appears to offer unlimited scope for the resident engineer to change whatever he pleases. However, the accepted interpretation of its meaning places significant restrictions on what constitutes a valid variation. A fundamental point is that a variation is not some special remedy for the contractor's failure to execute the works properly, but a change which is considered necessary or desirable even though the contractor is performing his obligations satisfactorily.

Clause 51 does not define a variation exhaustively, but states that it may include 'additions, omissions, substitutions, alterations, changes in quality, form, character, kind, position, dimension, level or line and changes in the specified sequence, method or timing of construction'. This may appear exhaustive enough for most tastes but the list, although comprehensive, is qualified by the context of the clause. Variations can be ordered to 'any part of the Works' when they are needed for the 'completion of the Works' or its 'satisfactory functioning'. It is generally accepted that these words mean that changes must be restricted to parts or sections of the works only and cannot be on such a scale as to alter the character of the project as a whole. This applies not only to additions or modifications but also to omissions, as major deletions are unlikely to assist the completion or functioning of the works as described in the contract documents.

Furthermore, for a variation to be valid, it must be 'necessary for the completion of the Works', or, if issued for any other reason, it must at least be 'necessary or desirable for the satisfactory completion and functioning of the Works'.

This would seem to exclude changes made simply because the employer is short of money or has changed his mind about the kind of job he wishes to have built. Accordingly, every variation must be a response to circumstances arising or becoming known during the course of the contract, and by its execution allows the works to be completed, or to function after completion, in the manner originally intended. Thus a variation occasioned by the discovery of a layer of soft material in a foundation is valid, as is one to correct a design fault but an order to delete one of the settling tanks from a sewage treatment plant as an economy measure would be invalid. This view is not universally accepted and some discern a much wider freedom to vary the works, indeed, one judge considered that a clause similar to clause 51 could have been used to order the contractor 'to erect a series of buildings on the site, pull them down, and put them up again ... as often and as long as they chose'.

It follows, if the limited interpretation is adopted, that a variation must relate to something in the contract documents, even if it covers additional work there should be some evidence to demonstrate why it is needed. Variations cannot be issued in respect of the contractor's temporary works (unless, of course, particular temporary works are specified in the contract documents) and the resident engineer must use other powers if he feels it necessary to intervene (see Chapter 5). Similarly, when the agent submits a proposal for approval, such as his programme, a method of working, or a source of materials, which is entirely at his own discretion and not specified, and the resident engineer subsequently changes his mind then this is not a variation. If, however, the contract documents contain a list of 'deemed-to-satisfy' alternatives from which the contractor can select, any change ordered by the resident engineer would be a variation as it eliminates the choice offered to the contractor at tender stage.

In contrast to the normal ban on the resident engineer interfering in operational and programming matters, clause 51 allows him to issue variations in respect of 'sequence, method or timing of construction', but only when these are 'specified', that is, when they are set out in the contract documents. Thus, if sectional completion is incorporated in the contract, a variation could be issued to change the order or timing of the various stages. Similarly, if a method of construction is detailed in the specification, a variation could be ordered to change it in whole or in part. This part of the clause, like that covering changes to the design, cannot be extended to cover the whole of the works and cannot be used, say, to bring forward the contract completion date. Where operational details are not included in the contract, the resident engineer cannot intervene except through the 'satisfaction' clause (clause 13 in the ICE Conditions of Contract) or the provisions relating to the approval of the contractor's programme and method of working (clause 14). When clause 51 applies, the provisions for valuing a variation allow the resident engineer to take appropriate bill rates into account, otherwise the contractor's actual extra costs must be reimbursed.

When the contract documents are incomplete and late drawings have to be issued in accordance with clause 7 of the ICE Conditions of Contract, this action is considered to be equivalent to the ordering of a variation and is subject to the same procedures.

A variation order may result in a change of quantities, but the converse does not apply. Changes arising from minor inaccuracies in the contract documents, the tolerances normally associated with averaging or the difficulties in assessing quantities in rough, inaccessible or tree-covered terrain do not require ordered variations.

What happens if the resident engineer oversteps his authority? The contractor is entitled to assume, unless notified otherwise, that a resident engineer who is empowered to issue variation orders can do so without any cost limit. The employer will be bound by the variation and will have to pay for its execution. If the problem is not one of authorisation, but of an ordered variation which falls outside the contractual meaning of the term, the contractor can seek formal clarification from the Engineer. If it is given, he has two choices: he can either refuse to proceed, although this is a risky option because if his interpretation of the contract is wrong he may find himself in breach or liable for liquidated damages; or he can obey the order under protest and submit a claim for the extra cost and delay. It is important in such a case that the contractor acts at the first opportunity, he cannot lead the employer into a trap by acting on an order, accepting payment for it as a variation and then claiming at a later and more convenient time.

Finally, it should be noted that the employer has no power under the contract to order any variation on his own initiative. All variations must be issued by the Engineer for the works or his representative.

Issuing a Variation

Clause 51(2) states that a variation order must be in writing. The importance of ensuring that the resident engineer's exact requirements are properly stated and clearly understood by the contractor make this an obvious stipulation, as does the need to have a permanent record of the details of every change to permit accurate evaluation.

In common with all site communications (see Chapter 8), a variation order must be written in clear English, be dated and contain every detail necessary to identify what it covers (whether expressed in words or shown as a drawing or both). The writer must avoid superfluous comments or observations which generate confusion at the least and unjustifiable expense at worst. If an agent finds unnecessary additions to the basic information, it is understandable that he will seek to explain their inclusion and predictable that his interpretation will tend to be favourable to the contractor. It is in this fertile ground that many disputes flourish. Variation orders should be terse and, like site letters, short on pleasantries: 'please' is, in contractual terms, an unnecessary word.

To be effective, a variation order has to cover the following essential details:

(1) a concise description identifying the subject of the variation, including the item number from the bill of quantities where applicable;
(2) the location in the works, cross-referenced to any drawing on which the particular item may be shown;

(3) a clear statement of the nature of the change expressing it as an omission, addition or modification;
(4) details quantifying the extent of the change, given precisely where possible but otherwise clearly described as an approximation or estimate;
(5) details of timing, but only when the variation is in respect of specified operational matters or when it is essential to incorporate restrictions of this kind into the variation (the general rule is that a variation order tells the contractor what to do but not when to do it);
(6) any other information which a contractor might reasonably require to execute the variation order to the resident engineer's satisfaction (for example, a stipulation that a member of the site team must be present when the work is carried out);
(7) a statement of how the resident engineer proposes to value the variation, indicating whether the normal process involving consultation based on bill rates is to be used or daywork rates applied.

The pitfalls in writing variation orders can be seen in the two examples which follow. First, the unclear and incomplete order:

'Please arrange for all 35mm L-shaped bars in the wing walls of the Railway Bridge to be cut to 800mm length on the short arm (Previous instructions are superseded).

6 No. additional bars are also required for the NW wing wall, to be placed as shown in the reinforcement drawing. This work is to be done first to meet the track possession dates for erecting the shuttering at this wing wall.'

The writer knew what he meant, but he would have made it easier for the agent organising the work, the inspector supervising it and the site staff carrying out the valuation if he had been more precise and comprehensive:

'Reinforcing Steel—Wing Walls to Structure 6 (Railway Bridge)

(a) The following varied work is to be carried out. The 66 No. L-shaped reinforcing bars (Mark 45 on page 13349/05 of the Bending Schedule) are to be re-cut on site to provide a length of 800mm on the short arm instead of 950mm as scheduled. (The variation issued verballyat 11.00 a.m. today stating a length of 750mm is superseded).

(b) The following additional work is to be carried out. 6 No. L-shaped reinforcing bars to Mark 45 as modified in (a) above are to be supplied for the North West wing wall of Structure No. 6. Position is correctly shown on Drawing 1108/36/2W but the number of bars is incorrectly stated on the Drawing and in the Bending Schedule as 14 instead of 20.

(The Contractor's attention in drawn to the need to complete the North West wing wall in accordance with the agreed track possession schedule.)

Payment for (a) at daywork rates, for (b) at bill rates.'

The second example illustrates the order filled with irrelevant and confusing detail:

'Stone-filled batter drains and V-ditches, Beech Farm Cutting

The 150mm diameter stone-filled batter drains shown on the east and west side slopes of Beech Farm Cutting between chainages 2100 and 2900 (see Drawings 5824/24/2W and 5824/25/2W) not required due to the lack of seepage from the sandy layers exposed in the side slopes. In order to prevent erosion by surface run-off from the adjacent land they are to be replaced by extending the standard unlined V-ditches at the top of each of the batters from ch.2100 southwards for about 700m. My Section Engineer (Roads) will determine the final limit of the ditches by a site inspection and should be contacted at the appropriate time.

I consider this additional ditching to be similar to the work priced in the Bill of Quantities and propose to value the variation accordingly as provided for in clause 52(1).

Note that the 1.5m offset shown on Drawing 473/D4/1W is measured from the top of batter to the near edge of the ditch and not the centre-line - recent work has been done incorrectly and your proposals are awaited.'

This writer has included unnecessary information on ground conditions (which might be inconsistent with details in the contract documents and thus detrimental to the employer's interests), offered two reasons for the presence of batter drains (when no explanation was required) and given an unreasonably vague indication of the extent of the additional work. The details of payment do not make it clear whether the additional ditches are to be valued at bill rates or at a new rate based on that in the bill: both options are covered by clause 52(1) and the agent could interprete the wording as an offer to use the second method although the resident engineer intends to use the first. The note on the setting out error is irrelevant to the varied work and can only serve to confuse the issue by getting a straightforward rejection of improper work mixed up with a variation. A better course of action would be to write a letter rejecting the incorrect ditching and to issue a simple variation order:

'Drainage—Beech Farm Cutting

(a) The following work is to be deleted.
Item 3/13/A (stone-filled batter drain), chainage 2100-2900, east and west sides (see Drawings 5824/24/2W and 5824/25/2W).
Total quantity: 560 linear metres.

(b) The following additional work is to be carried out.
Excavate standard unlined V-ditch (Item 3/14/C, standard detail as Drawing 473/D4/1W) from chainage 2100 to 2800 east and west sides, to extend proposed cut-off ditch shown terminating at chainage 2100 on Drawing 5824/24/2W.
Excavation beyond chainage 2800 to be determined by site inspection -contact Section Engineer (Roads).
Total quantity: approx. 1400 linear metres (subject to confirmation).
Payment to be at Bill rates.'

The ICE Conditions of Contract recognise that there are times on a construction project when variation orders are initiated out on the site, over the telephone or at a meeting. In such circumstances the resident engineer must act as quickly as possible to provide the agent with a written confirmation, however, clause 51(2) provides the machinery for the agent to produce his own confirmation which, if not corrected or repudiated by the resident engineer, becomes effective as a variation order itself. These confirmations (usually known as 'confirmations of verbal instructions' or CVIs) should be studied carefully for any inaccuracies, differences of emphasis or additional content. Any discrepancy, however trivial it may seem, must be noted and clarified. It is good practice, albeit unnecessary contractually, to issue a formal order even if the CVI is correct in every respect as this keeps up a full and consistent record of all variations.

The possibilities for differences between the contents of a CVI and the resident engineer's intended variation order are endless. For example, compare this CVI with the variation order which follows.

'CVI No. 38
The face of the concrete retaining wall at Car Park 2 is to be lightly bush-hammered to expose the aggregate.
Payment to be made on agreed plant and labour returns.'

'Variation Order 276
The following additional work is to be carried out.

Subject to a satisfactory 1.5 × 1.5m trial panel (location to be agreed on site), the face of the northern boundary wall at Car Park 2 is to be lightly bush-hammered to expose the aggregate. The bush-hammering is to extend from 100mm above finished ground level to 50mm below the top edge chamfer. The eastern boundary wall is not to be treated without a further order.
Payment to be based on Bill Rate for Item 10/4/E (bush hammering walls at Main Access Ramp); plant and labour returns may be submitted to assist in valuation.'

The resident engineer must ensure that every variation ordered in accordance with clause 51 is clearly identified as such. Variations must not be included obscurely in letters covering related topics nor expressed in terms which make their status ambiguous. The practice of having a special form on which to issue variation orders is a most effective means of drawing attention to their special nature but to make the most of this system it has to be operated strictly and consistently.

Instructions and variations
The resident engineer has the authority to issue instructions to the contractor on a number of matters. He can order him to remove improper work or materials, expose work which has been covered up without permission, re-execute unsatisfactory workmanship and in general to carry out the requirements of the contract. None of these are variations and the agent must be left in no doubt on that score. If the normal method of issuing variation orders is by letter, then the heading or opening sentence must make the status of the communication clear by including the phrase 'ordered variation' or 'variation order' or by referring to clause 51 of the ICE Conditions of Contract. Similarly, any instruction must contain a reference to the clause from which the resident engineer draws his authority, to exclude the possibility that the agent might mistake it for a variation.

If special forms, often printed on coloured paper, are available, their use must be properly controlled, otherwise the benefits are rapidly replaced by confusion. Site staff begin to see the special form as representing some superior channel of communication, carrying greater weight than an 'ordinary' letter and very effective at getting a reluctant agent to act. Its effectiveness is invariably an illusion.

Whatever the reason, the result of any ambiguity between an instruction and a variation is all too predictable. The site staff, satisfied that they have stated their requirements and given the agent his instruction, rest content in the knowledge that the work will be done, all at no extra cost to the employer. The agent, on the other hand, accepting what he reads as a variation, sets out to do the work, secure in the belief that he will be paid for his efforts. A period of mutual satisfaction follows which, even if neither the resident engineer nor the agent voice any doubts, is inevitably short-lived. When the next monthly certificate is processed the resident engineer will be irritated to see an extra included for carrying out an instruction to work to the contract and the agent will be correspondingly annoyed when his claim for payment for a variation is struck out.

Bad feeling and an erosion of the co-operation essential to the smooth running of a construction project are the likely outcome of these failures in communication. Clarity in the expression of his requirements, whether instructions or variations, is the responsibility of the resident engineer and the blame for any confusion must ultimately rest with him.

A contractor was employed to carry out the reconstruction of certain harbour facilities. As part of the works he was required to refurbish some existing reinforced concrete hardstandings, planing the surfaces to provide uniform crossfalls and cutting grooves to give a non-skid texture. The techniques of planing and texturing were not specified, and the contractor had great difficulty with his chosen method. A site meeting was called. The assistant resident engineer attended to point out that the contractor's method was producing an unsatisfactory result. The agent's intention was to persuade the assistant resident engineer that a variation order was required.

The assistant resident engineer drafted a letter to record the discussion:

'I refer to the site meeting on 21 September at which it was agreed that the grooving work to the concrete aprons at Loading Bays 1–6 was not of an acceptable standard.

I confirm that the current method of working has not produced the required result and that your proposed new method, involving the use of heavier equipment and a preliminary cut followed by the main cut is acceptable and is to be employed on all the remaining areas. The grooves in the aprons already treated are to be given a second cut using the new equipment.

Mutually agreed plant and labour records are to be kept in respect of the operation.'

This draft lacked a firm statement of the contractual position and would have encouraged the agent to construe it, innocently or otherwise, as confirming the variation order he was seeking, with payment to be based on plant and labour returns. A second version was prepared which contained clear instructions and a proper statement on liability. Even the most optimistic Agent could not read it as a variation order:

'I refer to the site meeting on 21 September regarding the groove cutting on the concrete aprons at Loading Bays 1–6. I confirm that the work so far completed does not comply with the specification and cannot be accepted into the Permanent Works. I consider that the equipment you have chosen to employ is inadequate and approval for its use is withdrawn. In accordance with clause 14 of the Conditions of Contract I requested your new proposals and confirm that the use of heavier equipment to cut the grooves in two stages is approved.

In accordance with clause 39 I instruct you to re-execute the groove cutting completed to date and confirm that I accept your proposed method, namely to repeat the cut along each groove using the new heavier equipment.

All the above is considered to be covered by your contractual obligations and is to be undertaken at your own expense.

I note your request that mutually agreed plant and labour records be kept in respect of this operation and, without prejudice to the statements on liability above, have no objection to such records being maintained.'

Valuing a variation

The procedure for valuing a variation is set out in the first two sub-clauses of clause 52 of the ICE Conditions of Contract. Although these sub-clauses contain subtle differences in their wording, for instance, in the first the rate or price is 'determined', whilst in the second it is 'fixed', they nevertheless appear to repeat each other. It is generally accepted that clause 52(1) covers the majority of additional or modified work. Clause 52(2), however, is considered applicable in only two particular cases: the valuation of omissions and the assessment of those variations in which the payment of the full bill rate for extra work would provide the contractor with excessive profit which must therefore be 'clawed back'. This latter case occurs when a proportion of the bill rate covers fixed overheads which do not increase pro rata with the quantity of work. For example, if the construction, maintenance and removal of a temporary diversion is priced in the bill at a rate per week, then doubling the length of time for which the diversion is required does not double the cost to the contractor and the rate is adjusted accordingly.

Once a variation order has been issued, the resident engineer has a duty to value its effect. Three results are possible: no cost; cost saving; and extra cost. It can happen, and there is certainly nothing in the contract to prevent it, that a modification produces no delay, disruption or additional work and should fairly be valued at zero. For example, a minor change in levels notified well in advance would fall into this category, as would the substitution of one type of fencing for another of the same cost, requiring no extra labour and ordered before the contractor had made any purchases from his supplier. In many cases, however, variation orders finally valued at 'no cost to the Employer' are so assessed because they cover work considered to be part of the contractor's obligations, that is, they were instructions which were incorrectly issued as variations.

When the contractor is ordered to omit work there is a presumption that a reduction in the contract sum can be calculated on the basis of the bill rate for the item and the quantity deleted. This is not always the right course, however, as leaving out work, whilst simple enough for the resident engineer, can cause many problems for the agent. Orders may have been placed for materials subject to substantial cancellation charges, delivery may already have been taken and payments may have been made. Plant or other resources may have been assigned to the site and will still incur full costs. Preparatory work or temporary works may have been carried out in anticipation and other operations which may have been programmed to fit around the cancelled work could possibly have been more economically organised if the agent had known of the omission earlier. The contractor will be entitled to payment. The contractor may further argue that the extent of the omission is so significant that the bid submitted at tender stage is now completely unbalanced and seek to claim loss of profit. In fact, neither clause 52(1) nor clause 52(2) offer any remedy to the contractor. He will have to show that the change amounts to a breach of contract and sue for damages.

Variation orders for modified, new or additional work involving extra cost are the most common, and the procedure set out in clause 52(1) is used to evaluate them. Where the variation increases the quantity of an item already in the bill (for example, instead of 25 lighting columns along a service road, the spacing is changed and the contractor ordered to provide and erect a total of 30), the contract rate can be used in valuation as the work is 'of similar character and executed under similar conditions'. However, when the work is not directly comparable to that priced in the bill (for instance, if the five extra lighting columns are to be erected in another part of the site where access, excavation conditions and working space are different from those on the service road), the contract rate or price may be used 'as the basis for valuation as far as may be reasonable'. This process requires consultation between the site staff and the contractor involving, for instance, the comparison of billed work with the varied work or the analysis of the relevant rates to permit a new rate to be built up. The co-operation of the agent is essential, but, as long as the resident engineer and the agent are agreed on the principle of using the bill rates as the basis of evaluation, any dispute is likely to be centred on the cost differential in terms of plant, labour and materials and these should be a matter of record. The two fundamental requirements, therefore, are: acceptance of the principle of 'analogous rates'; and a set of mutually-agreed returns covering the varied work.

It may be that the nature of the work ordered, or the extent of the variation, is such that no appropriate bill rate exists. In such cases the ICE Conditions of Contract require the resident engineer and agent to consult in order to agree a 'fair valuation', which is generally accepted as the reasonable cost of the work, plus a percentage for profit. The assessment of what is reasonable, for both works cost and profit, must be objective and justifiable as the settlement which any experienced resident engineer and agent would agree upon.

The process of consultation set out in the ICE Conditions of Contract first considers the direct application of bill rates, then their indirect application and finally the establishment of a 'fair' price. They do not define how this consultation is to be carried out nor what it should cover, but clearly the resident engineer must give the agent the opportunity to state his case and, if he considers the bill rate to be inappropriate, to present his proposals as to how the rate should be adjusted or what basis should be used to assess a new price. For his part, the agent must support his submission with full particulars including, where applicable, information on the build-up of his contract rates.

In the majority of cases the leading role at this stage must be taken by the agent, for he is, or ought to be, in possession of all the appropriate time sheets, invoices, accounts and pricing data. The resident engineer generally confines himself to checking the accuracy of these particulars and assessing their relevance to the valuation. The process must not be seen as a trial in which one party carries the 'burden of proof', for in an exceptional case the resident engineer can exercise his discretion and pay the contractor the reasonable value of the work even when costs cannot be fully substantiated. The agent must nevertheless take the initiative in establishing the case for extra payment.

Where it is reasonable to value at the bill rate or some modification of it, the fact that the tendered price may be wrong or deliberately set low is irrelevant. Although this may be the case, it cannot alter the principle on which the valuation process is based, namely, that the 'character and ... conditions' of the work are the relevant factors which have to be taken into account. The contractor has agreed to execute all the work in the contract, both original and varied, at the rates in the bill, and it would destroy the basis of competitive tendering if, having won the work, he were subsequently allowed to avoid the consequences of offering certain loss-making rates as part of his successful bid.

Failing any consensus on how the valuation should be made, the resident engineer must settle the matter himself. If the consultation has been conducted in an atmosphere of co-operation, all the information is available for the valuation to be made and, if the contractor wishes, taken on for final settlement as a contractual dispute. When, however, the agent has not been helpful, refusing, for instance, to submit full particulars in support of the price he is seeking or failing to disclose what that price might be, the resident engineer is under no obligation to put up his own detailed calculations for the agent to criticise. The ICE Conditions of Contract require the price or rate to be determined after consultation and the contractor notified. There is no requirement to provide any statement as to how the value has been calculated. It can be appreciated that, if the resident engineer had to produce such information, an unscrupulous contractor might encourage his agent to take no part in the consultation process and thus force the resident engineer to set a 'floor' to the valuation which could then be disputed and bid up towards a 'ceiling' more advantageous to the contractor. This would substitute a kind of haggling for the proper method of valuation.

Similarly, the resident engineer must ignore any inducement to enforce a valuation favourable to the employer. Pressure from this direction need not be direct, for in those cases where the site staff are in the full-time service of the employer there may be a desire, albeit subconscious, to save money for the parent organisation which can result in a bias against the contractor. Any tendency to favour the employer, however caused and no matter how well-intentioned, must be recognised and resisted.

The major problem in valuing any variation lies in assessing the extent of its indirect effects. Executing work out of sequence, on a different scale to that specified in the contract documents or in a different manner to that originally envisaged is likely to produce costs over and above those directly attributed to the operation itself because of the effect on other adjacent or concurrent activities. Indeed, work which is remote in distance and time can still be influenced by a variation because of some operational link or chain of cause and effect.

A variation order was issued on a brewery contract to extend a deep drainage run. The extension crossed the only all-weather access to one section of the works. Consequently, deliveries of materials into this area were erratic, reducing the output of the men and equipment working there to an uneconomic level over a two-week period.

The Contractor building a sewage outfall was ordered to change the foundation level of a concrete slab after the excavation had been completed and trimmed. The only suitable machine present on site had to travel from the far end of the works across an area which had been badly rutted by construction traffic. Although the re-trimming operation took only a few hours, the machine and its operator were not available for any other productive work for the whole day.

The contract documents for an airport extension specified a sequence for the construction of a series of mass concrete walls. Before work had commenced the resident engineer issued a variation order bringing forward some of the sections of wall. The agent demonstrated that the original schedule allowed much greater scope for re-using formwork than the revised version and argued that the bill rate should be increased to cover the extra cost of labour and materials required to make the additional shutters.

Late in the construction of a large river bridge the contractor was ordered to bring in a large quantity of additional imported fill. The only available haul route ran along a section of partly completed carriageway on which the sub-base layer had been placed. The additional traffic disturbed this layer which had to be re-trimmed. It also caused the loss of a significant amount of sub-base as a result of penetration into the sub-grade, contamination with spilled material from the lorries and displacement off the carriageway. The agent sought a 'fair valuation' for these indirect costs.

The valuing of variations is rarely straightforward and the resident engineer must, within reason, take into account all their side-effects. 'Indirect losses', 'delay and disruption', 'consequential costs' are all phrases used to describe, somewhat antiseptically, the financial consequences which may legitimately be passed on to the employer. 'Botheration factor' conveys the meaning with greater accuracy, although the vernacular of construction sites contains a more robust (though similar sounding) term.

In making his case for an enhanced rate or a 'cost plus' valuation, the agent must set out the botheration element honestly. It is both unwise and unnecessary to resort to invention, adding on a few hours of machine time to the record sheets, including for a non-existent banksman with an excavator, or submitting returns for resources and materials which were never used. Fictions such as these are open to discovery at final account and show a lack of respect for the resident engineer's capacity to determine and support his own judgement. Site staff should never connive in any falsification of records for it damages their own authority and undermines the vital principle of accurate and reliable record-keeping. If equipment could only work at 50 per cent efficiency then this should be stated and the resident engineer's confirmation invited. If operations elsewhere are delayed or left short of supplies, then the agent's estimate of the resultant loss should be submitted for the resident engineer's inspection and assessment. General botheration is not a 'grey area' which needs to be dealt with by sleight of hand or by 'cooking the books'. Properly explained and supported it is a justifiable charge against a variation and neither the agent nor the resident engineer should have any qualms about treating it openly.

Daywork

The resident engineer has the option of ordering that any variation be paid for at daywork rates. Clause 52(3) gives complete discretion to the resident engineer to take this course whenever 'in his opinion it is necessary or desirable'. The only limitation appears to arise from the wording of the clause which indicates that the order to use dayworks must be given in advance, but there is no other restriction and dayworks may be ordered even when rates exist in the bill which could be applied to the varied work. The contractor has no right to insist on the use of daywork rates to value a particular variation; similarly, he cannot resist the application of dayworks in lieu of bill rates although the result may be to his disadvantage. The only recourse for the contractor in these circumstances is the disputes procedure.

It is relatively rare for the employer to include a comprehensive dayworks schedule in the bill of quantities to allow tenderers to price the full range of items covering plant, skilled trades and labour. More typically, the dayworks section of the bill contains three provisional sums:

(1) for labour, to be charged at the basic rates of pay, together with the overtime rates, bonuses and 'plus rates' for skilled

trades, as set out in the 'Working Rule Agreement' of the Civil Engineering Construction Conciliation Board (the main negotiating body for wages in the industry);

(2) for materials, to be charged at the net price paid by the Contractor;

(3) for plant, to be charged at the rates laid down in the 'Schedules of Dayworks Carried Out Incidental to Contract Work' published by the Federation of Civil Engineering Contractors.

The employer inserts an estimated value for each, and the tenderers state a percentage oncost against the three items to cover all their overheads, charges and profits. Together they make up the dayworks element in the contract sum. Usually, the materials and plant items carry only small percentages whilst the figure against the labour item is very large, often well over 100 per cent, reflecting not only the high overheads borne by this element of the contractor's resources but also the indirect cost of diverting labour to unprogrammed work.

Generally, contractors ensure that their percentages are pitched at the level where dayworks is attractive to them as a means of valuing most variations. Agents are keen, therefore, to suggest dayworks as the basis for valuation and are sometimes reluctant to proceed with any varied work not closely identified with a bill item unless they are given an assurance that daywork rates will be paid. There is, of course, no contractual justification for such a stance and if the resident engineer decides against dayworks the agent must proceed with the work as ordered, relying on the 'fair valuation' process.

The advantage of daywork rates is that their use removes the uncertainty over valuation. This can be useful to the resident engineer as well as the agent, as a contractor already satisfied with the method of payment is likely to co-operate more fully in the execution of an awkward variation, or one involving unpredictable factors. In addition, a great deal of expensive staff time can be absorbed by long wrangles over valuations, and payment on dayworks narrows the area of potential dispute into the relatively simple question of records.

Clause 52(3) lays down very strict procedural requirements for the agent and his staff to follow:

(1) All receipts, vouchers, etc. must be kept for inspection.
(2) Quotations must be obtained for all materials and submitted for approval before ordering.
(3) A full record of all plant and labour (including names of workmen) employed in the daywork operation, together with descriptions and quantities of any materials used, must be submitted daily in duplicate for the resident engineer to check.
(4) A priced statement of plant, labour and materials must be presented each month.

If any of these items is submitted late or omitted, the contractor loses entitlement to payment. The resident engineer is given a discretionary power, when he considers that the proper presentation of the required details is impracticable, to make his own assessment at daywork rates or to revert to a 'fair and reasonable' valuation based on bill rates or any other suitable method, but the onus, in the first instance, is on the contractor to provide the specified information. This places a heavy burden on the agent, for the timetable is a difficult one to meet particularly when it is remembered that the work in question, is a change in the contract and will not have been programmed.

The submission of quotations for materials in advance of ordering can be a useful safeguard when very expensive components are involved or when more than one acceptable source exists, and it does give the resident engineer an indication of whether the cost of the variation is likely to be excessive or outside his authority. When a variation order is issued because the change is considered 'desirable for the satisfactory completion and functioning of the Works' such considerations are valid. If, however, the variation arises because the resident engineer believes it is 'in his opinion ...necessary for the completion of the Works', it would seem imprudent subsequently to withdraw it on the grounds of cost, indeed, the wording of clause 51 ('shall order') implies that the resident engineer has a duty in such a situation to proceed with the issue of the appropriate variation. Under normal circumstances, the site staff should be able to form a reasonably accurate estimate of the materials cost in a proposed variation, and the 'fair valuation' provision protects against the contractor charging an exorbitant price. When a variation order is urgent it is in nobody's interest to wait for formal quotations to be obtained as there is every likelihood that the cost of the damage, abortive work or standing time which results from the delay will exceed any benefit derived.

The presentation of daily records is a practice honoured more in the breach than by observance. The staff in a busy agent's office are generally hard pressed by their routine work and find it difficult to extract the information, let alone write it up on the contractor's standard form and despatch it, in duplicate, to the resident engineer to be on his desk the next morning. Provided the site staff are given a clear indication of what resources the agent is committing to the daywork operation, records delivered within 24 hours of the work to which they refer can be considered an acceptable discharge of the obligation, as memories are still likely to be fresh and discrepancies still easily resolved by a site inspection. Some further flexibility is reasonable when, for example, the weekend intervenes, or other justifiable reasons exist to explain a slow delivery. There is, in practice, no definite point at which a line can be drawn to exclude late submissions. Certainly records a week old are unacceptable except as a basis for a discretionary 'fair valuation', but presentation after 36, 48 or even 72 hours may be considered reasonable in particular circumstances.

When work is executed at daywork rates, the resident engineer is effectively 'hiring-in' the contractor's plant and labour at, in most cases, hourly or daily rates. The cost of any inefficiency, wasted time or lack of full effort is now borne by the employer, in contrast to the situation when work is done at bill rates, or rates based on the bill. The resident engineer must take a close interest in the operational control of all daywork activities and the site team must be prepared to intervene, in a way which would be quite improper for work at bill rates, in any instance of uneconomic working or unnecessary use of resources.

A contractor working on a motorway interchange was ordered to demolish an old boundary wall, shown on the drawings as being unaffected by the works. The proximity of a school playing field meant that the operation would have to be carried out under severe restrictions on working hours and means of access and that special precautions would have to be taken to ensure safety and security were maintained. No bill rate existed for any sort of brickwork demolition. The resident engineer ordered that daywork rates be applied.

The drainage work on a contract to build a new industrial estate was executed by three specialised gangs which left the site after this operation was completed. Much later the resident engineer warned the agent that he would be issuing a variation order for the laying of some additional drains. The agent could not recall any of his original gangs and pointed out that the work was not only out of sequence but on a much smaller scale than that priced in the bill. It was agreed that the varied work would be done using resources available on the site and the order was issued with daywork rates specified for valuation.

A contractor building a viaduct was ordered to undertake additional earthworks and, because the operation was in a restricted part of the site and on a limited scale, the resident engineer notified the agent that daywork rates would be applied. During the progress of the work the inspector supervising it became concerned at the slow output and found that two of the four motorscrapers being used were not fully operational. The resident engineer instructed the agent to remove them from the dayworks operation.

Prime cost items and provisional sums

The bill of quantities may contain prime cost (PC) items and provisional sums (see Chapter 4). The powers and processes set out in clauses 51 and 52 in respect of ordered variations also apply here.

Prime cost items are usually executed by nominated subcontractors. The resident engineer issues an order requiring the contractor to enter into an agreement with the firm named in the contract documents and the work is then carried out with the contractor receiving a lump sum or a percentage for his 'attendance'. However, under clause 59A the contractor can object to the nomination if he has reasonable grounds for doing so: if, for example, the nominated sub-contractor refuses to take out normal insurance cover, or if the main contractor doubts the financial stability of the sub-contractor. In such a case the resident engineer has four options available to him:

(1) instruct the main contractor to proceed despite his objection, subject to the employer covering such losses as may arise as a consequence;
(2) make an alternative nomination;
(3) issue a variation order to omit the PC item from the contract (the work may then be done under a separate agreement between the employer and another contractor, by the employer's own workmen or not at all) and pay to the contractor the sum in the bill for attendance;
(4) issue an order for the main contractor to do the work.

The last option is always open to the resident engineer, for clause 58 gives him the power to order the contractor to execute any work covered by a PC item, regardless of whether a sub-contractor is nominated in the contract documents. However, the clause contains a curious ambiguity for the 'power' is subject to the contractor's consent, and is thus no power at all. A contractor might reasonably refuse such an order when the work is of a specialised nature and outside his field of expertise, but the wording appears to give complete and unfettered freedom to an agent to turn down the PC item on any grounds. It is possible that a term could be implied in the contract to the effect that the contractor's consent would not be unreasonably witheld, but this is by no means certain. The agent is given the whip hand in negotiation, for payment is either on a quotation accepted by the resident engineer or by the normal clause 52 process for valuing a variation. Thus the agent could set a price higher than he might expect to get under a clause 52 valuation and invite the resident engineer to 'take it or leave it'. The moral here is not to use PC items for any work where the main contractor is likely to be directly involved, and to ensure that nominations of sub-contractors are unlikely to fail.

Provisional sums, unlike PC items, are contingent and the contractor cannot claim for any losses if they are not used. Neither does the contractor have any right to refuse an order to do work covered by a provisional sum. Clause 58 authorises the resident engineer to order the main contractor to execute the work and also allows him the option to nominate a sub-contractor to do all or part of the work if he so desires, payment then being in accordance with clause 52.

Chapter 11
Measurement

The resident engineer and his team are predominantly concerned with the 'engineering' aspects of the contract: setting-out; the contractor's programme and methods of working; acceptability of materials and quality of workmanship. Financial matters tend, in general, to be relegated to a secondary position and it often takes some dramatic development, the presentation of a major claim, for instance, to turn their attention to money. Even then the reaction is frequently indignation at the contractor's preoccupation with making money at the expense of 'getting on with the job'.

To many site staff the regular process of payment is seen, if not as a side issue, then as something incidental to the real business of building the works, whereas for the contractor and his agents these payments are the basic reason for their presence on site. Although it would be wrong for the resident engineer and his team to apply commercial values to their duties, this widespread failure to appreciate the business side of contracting can be a major barrier to understanding between the agent and those supervising his work.

The employer, too, is concerned about money and is entitled to a full account of how and why it is being spent. He is, of course, paying not only for the construction of the works but also for the supervision. In return the employer will expect to have the project completed satisfactorily and without unnecessary cost, and so the resident engineer must have the most comprehensive and accurate information possible to ensure that all liabilities and payments can be properly determined.

Paying for the work

Contracts under the ICE Conditions of Contract are measure-and-value contracts. The resident engineer has the duty of determining by 'admeasurement—a traditional term embracing subjective assessment by means of professional judgement as well as straightforward enumeration and measurement by dimensions—the value of the contractor's work.

All contracts specify procedures for payment. In the ICE Conditions of Contract, clause 60 sets out the method by which the contractor receives interim monthly payments during the course of the works and obtains settlement of the final account on completion. The contractor is obliged to submit a monthly statement which, unless any special requirements are laid down elsewhere in the contract documents, must contain the following details:

(1) the estimated value of the permanent works completed up to the date of the statement;
(2) the value of any materials delivered to the site but not yet incorporated in the works (or, if stored off-site, made the property of the employer);
(3) the amount of any extras or claims to which the contractor considers himself entitled.

The first part of this three-part submission follows the layout of the bill of quantities. The contractor inserts his estimate of the cumulative quantities to date against the various items and extends them by the bill rate to produce the interim values, which are carried forward to a summary. In the second part, the contractor is seeking payment for goods or materials, the cost of which usually forms part of a bill rate along with other elements such as labour. As these materials are still to be incorporated in the works the full rate is not payable (except when the appropriate item in the bill covers only the supply of particular components) and the submission takes the form of evidence of their value, delivered to site. If the goods or materials are stored off-site, the contractor must show that they have been transferred into the ownership of the employer by means of a legal process known as vesting. In the third part, which covers all the 'extras', the contractor writes his own item descriptions and produces an evaluation which may be based on a quantity and a rate but is more likely to be a lump sum.

These formal submissions are supplemented by the contractor's calculations, dimension sheets and any other relevant information to show how the quantities or sums in the statement have been produced. In the case of claims for extra payment, this supporting information must be in sufficient detail to enable the resident engineer to investigate the claim. Even when the principle of additional payment is accepted, the contractor is entitled only to such sums as he can substantiate with 'full particulars'.

The monthly statement and its supplementary information is normally presented in triplicate: one file copy; one 'working copy' for the site staff to check; and one copy for return, amended as necessary, to the agent. The resident engineer examines the calculations, dimension sheets and other details, compares them with his site records and confirms or amends the figures presented in the formal statement. Any unacceptable or unfinished work is deleted and mathematical errors are corrected. When claims are involved, the first consideration is whether the employer has any contractual liability. If no basis for payment can be identified, the item is deleted and the agent notified accordingly. If the principle is accepted, the resident engineer assesses the value. Although an item may be rejected by the resident engineer, it is still worthwhile examining and commenting upon the supporting evidence to establish mutually agreed facts (or to determine the areas of disagreement) should the matter be taken further as a dispute.

Within 28 days of receipt of the contractor's statement the resident engineer has to examine it, correct it as necessary, and certify the interim payment to the employer, who must then pay the contractor. A typical interim payment certificate from a major highway scheme in the early stages of a two-year contract period is shown as Figure 11.

Lines 1—11 show the measurement proper, the resident engineer's evaluation of the contract work completed so far according to the various sections of the bill. The sums are all cumulative. Any amounts payable to nominated sub-

CONTRACT A34 SHAW BYPASS

Contractor X. Tipett

INTERIM PAYMENT CERTIFICATE NO. 7 FOR WORK

EXECUTED TO 30th SEPTEMBER 1982

			£		
Bill No.	1	General Preliminaries	319,924		1
	2	Roadworks General	444,188		
	3	Main Carriageway	265,043		
	4	Side Roads	2,050		
	7	Railway Bridge	137,938		5
	9	Footbridge (inc. payment of £33,000 to nominated sub-Contractor)	65,100		
	11	Underpass	39,887		
	12	Extension to Culvert	28,206		
	13	Retaining Walls	10,015		
	14	Dayworks	6,000	*	10
ADD		Balancing Item $\frac{1318371}{3955113}$ x £50,000	16,668		
		Measured Work	1,335,019		
		Extras	4,000	*	
		Contract Price Fluctuation	51,907		
			1,390,926		15
DEDUCT		Retention (5%)	69,548		
			1,321,378		
		Materials on site (at 97% value)	52,486		
		Vested materials	NIL		
			1,373,864		
DEDUCT		amount previously certified	1,124,108		20
		amount due	249,576		

Date statement received: 8.10.82 Date payment due: 5.11.82

* valued at current prices

I hereby certify that the sum of Two Hundred and Forty Nine Thousand Seven Hundred and Seventy Six Pounds is due to the Contractor

Signed M. Owen
Engineer's Representative

Date 29.2.84

Figure 11 Interim certificate. The line numbers on the right relate to the description in the text

contractors are shown separately (line 6), because the resident engineer has the right to demand proof that these sums are handed over without delay (ordinary sub-contractors do not enjoy this support and the timing of their payment is entirely a matter between them and the main contractor). Ordered variations which are valued at bill rates, or rates based on those in the bill, are normally included in the measure under the appropriate section of the bill. Work paid for as daywork is shown separately (line 10). The total value to date of all the bill sections is subject to an adjustment to take account of the balancing item, if the contractor has included one in his bid. The adjustment (positive or negative as the case may be) is proportional to the amount of contract work so far included in the measurement.

The sub-total thus produced (line 12) represents the value of the measured work, that is, work at bill rates and prices, executed up to the date stated in the heading to the certificate. When, as is the case in this example, there is a provisional sum in the bill for dayworks, the value of any dayworks in the certificate is included in the measured work section and the balancing item calculation. The dayworks figure, and the sum for 'extras' ('fair valuations' and claims) in line 13 are marked with an asterisk because they are at current prices rather than base-dated. The next figure, the contract price fluctuation (cpf) allowance (line 14) adjusts the base-dated measured work to current levels. Most contracts of over one year's duration have a fluctuation clause (see Chapter 1). The price fluctuation factor (pff), which must be worked out every month, is applied to the 'Effective Value' of the work in the certificate (that is, the amount of measured work completed since the last certificate) to determine the cpf allowance. The figure in the certificate is the cumulative total of the allowances to date. The index figures from which it is calculated are first published as estimates and later confirmed, so the sum is always subject to recalculation.

When payments in this section of the certificate are made against 'extras' in which the extent of liability or the sum involved has not finally been settled, they are often said to be 'on account'. In fact, all sums certified are 'on account', because the resident engineer can subsequently delete, correct or modify any element of an interim certificate, with the exception of payments to nominated sub-contractors, which cannot be reduced once they have been passed on.

The next sub-total (line 15) can be considered to represent the 'inflation-proofed' value of the permanent works completed so far. The resident engineer deducts from this sum a percentage known as the retention (line 16). Under the ICE Conditions of Contract the deduction on large projects is set at 5 per cent of the value of the works completed to date, subject to a maximum limit equivalent to 3 per cent of the tender total. The purpose of the retention is twofold: it provides a fund on which the employer can draw to pay for remedial work which the contractor may refuse to carry out and it acts as an incentive for the contractor to complete any unfinished work or repairs during the maintenance period. Half of the retention is released within 14 days of the issue of the certificate of completion and the second half becomes payable within 14 days of the expiry of the maintenance period, provided the contractor has done all the work required of him. The retention is held against contingencies and is deducted even when all the contractor's work to date has been accepted into the permanent works as satisfactory. Its purpose is not to cover the costs of rectifying defects or unacceptable work known to the resident engineer, such matters must be dealt with by making deductions or deletions from the contractor's monthly statement.

The resident engineer is empowered to include in his certificate (line 18) materials delivered to the site but not yet incorporated in the permanent works, fabricated steel sections, for instance, or pre-cast concrete units. The 'site' is, strictly speaking, the limits of the works as defined in the contract documents and so does not necessarily include the contractor's compound. Resident engineers who take a lenient view and certify in such cases are taking a risk. In the event of any breakdown in the contract, due to a serious breach, for example, or the bankruptcy of the contractor, the materials are not under their control and can be removed.

Goods and materials not incorporated in the permanent works are known legally as 'unfixed materials' and the most serious disadvantage in paying for them is that, in the event of the contractor's insolvency, someone with a better title than the employer may come along and repossess them. It may appear to the resident engineer that any materials which the employer has paid for are quite clearly his property, but it is only when materials are built into the works that they are firmly in the employer's ownership. In the case of 'unfixed materials' the legal principal of 'nemo dat quod non habet' applies, which in this context means 'you cannot sell what you do not own'. Thus if the contractor does not own the materials, because he has not paid his supplier for them, he cannot pass a good title to the employer although they may be included in a certificate and the employer may make a proper payment. Recent judgements have confirmed that if the contractor defaults the employer may find himself paying a second time to get possession of any unfixed materials.

The contractor took delivery of a large quantity of roofing materials and included them as materials on site in his montly statement. The resident engineer certified them for payment and the employer duly paid, although the contractor had not settled the account with his supplier. Shortly afterwards the contractor went into liquidation.

The supplier may have been only vaguely aware of the maxim 'nemo dat quod non habet', but he was perfectly familiar with the principle of 'the fast wagon', lorries were sent at once to the site with orders to recover the materials. The lorries were not fast enough, however, for on their arrival they found the site staff under instructions from the employer to refuse access to anyone seeking to remove goods or materials.

The supplier sued to repossess his property and won the case. The employer had to pay again for the materials and eventually used the supplier's labour to complete the roofing element of the original, but now abandoned, contract. (This is the case of Dawber Williamson Roofing Ltd *v* Humberside County Council (1979) 14 Build.L.R.70)

Paying for materials on site is a gamble only if the proper precautions are not observed. A prudent resident engineer will ask for proof that the supplier has been paid, a responsible contractor will not hesitate to provide it. The resident engineer has the discretion to value the materials at a figure he considers 'proper'. Usually this is the delivered price and so represents current levels, hence there is no cpf allowance and, because materials on site are specifically exempted by the ICE

Conditions of Contract, there is no retention. However, the contract does provide for the certified payment to be a percentage of the value of the materials. The figure to be used is stated in the Appendix to the Form of Tender (see Chapter 4), and the 97 per cent used in the example is typical.

Payment for materials on site can also be used to reimburse the contractor when the work covered by a bill item is only partly complete. If, for instance, a culvert is being constructed of pre-cast units which have been placed in position but not fixed together and sealed at the joints, the resident engineer may certify a sum based on the delivered value of the units.

The employer may decide to pay for certain materials which are not on site, using a vesting agreement to ensure, so far as is possible, that ownership is properly transferred. The materials and goods which will be treated in this way are listed in the Appendix to the Form of Tender and the procedure is covered by clause 54 of the ICE Conditions of Contract. On contracts where vesting is permitted, the interim certificates should indicate what is being paid. The certificate shown in the example was issued at a time when there were no vested materials. As with materials on site, the resident engineer values vested materials at the amount he considers 'proper'. This valuation is not subject to cpf adjustment nor to retention. The same percentage of the value is certified as for materials on site.

All the certifiable work, goods and materials has now been valued and adjusted according to the contract. Because cumulative figures have been used the amount due to the contractor is determined by deducting the total value of the previous certificate (line 20). The contract normally lays down a minimum value for payment on an interim certificate (the figure is stated in the Appendix to the Form of Tender) and if the amount due falls below this level the certificate is witheld and the next valuation is made on the basis of two months' work.

The timetable for valuation, certification and payment is strict and the 28 days is the time allowed for payment to be made to the contractor, not for the issue of the certificate. The resident engineer must ensure, therefore, that he has cleared the certificate for delivery, by the Engineer for the works if appropriate, to the employer well within this period, making a reasonable allowance for administrative procedures at the employer's office and postal delays. Late payments are subject to interest, usually calculated at a rate based on that charged by the major banks for commercial lending. Time starts to run as soon as the monthly statement is delivered to the resident engineer. The settlement date cannot be postponed because there are corrections or deletions to be made to the contractor's submission, unless they are so major, or unless the necessary documentation is so incomplete, that it can fairly be said that there is no valid monthly statement.

Interim certificates can be issued by the Engineer's representative if this power has been delegated to him. It is only the final certificate, issued after the end of the maintenance period, which must be signed by the Engineer for the works.

Agreed measurement

Interim certificates can be revised and because of the constraints imposed by the tight schedule, the measure is sometimes very much an estimate. Indeed, the practice of approximating two certificates out of three and doing the third more precisely is widely accepted in the industry.

This is satisfactory provided the necessity of eventually producing an accurate measurement for the final account is recognised. However, many items are covered up or rendered inaccessible at an early stage and their measurement cannot be deferred. Ordered variations, which are often less precisely described than bill items, must be measured accurately as they are executed. The preparation of the final account is itself tightly scheduled, and without systematically building up mutually agreed interim figures, the settlement of large items such as earthworks or structural concrete could not be achieved within the time allowed. In the investigation and assessment of disputed items the existence of reliable monthly quantities and measurement records can be of crucial importance. Certain items require no physical measurement. Lump sums for specific pieces of work, for example, the pumps in a sewage treatment plant or the pre-cast concrete beams in a bridge, are paid on completion, subject to compliance with the specification. Regular payments, such as those for maintaining diversions or servicing the site laboratory, are certified 'on demand', provided the resident engineer is satisfied that the associated contractual obligations have been fully discharged.

Nevertheless, the majority of work still requires measurement before certification, either because the estimated quantity in the bill must be checked against the amount actually carried out under the contract, or because the total quantity is executed over a period of months and the amount to be valued in a particular statement must be ascertained by inspection on site. Bulk earthworks is an example of an item which falls into both these categories.

The earthworks volumes quoted in the bill are based on ground levels which have to be checked after site clearance has removed any trees, buildings and other obstacles to accurate surveying. Further revisions are made by comparing the actual depth of topsoil strip against the quantity allowed for in the initial earthworks calculations. Changes in the division between suitable and unsuitable material and the issue of ordered variations modifying the side slopes of embankments or removing additional materials from cuttings are also taken into account. During this re-calculation any errors and inaccuracies detected in the original volumes are corrected, and the end-product is the complete re-measure of the earthworks elements of the bill by the time the final account is produced.

Because of the 'swings and roundabouts' effect inherent in the interpolation and extrapolation of levels, the final figure seldom differs to a great extent from the billed volume, but the difference can be significant in terms of the contractor's narrow profit margins. Certainly, no agent would accept as the final measure a figure based on the bill quantity adjusted only for variations.

At each monthly valuation when earthworks is in progress, interim quantities must be determined for inclusion in the certificate. By levelling at cross-sections in the areas where cutting or filling has taken place the current running total is determined either by manual calculation from the sections, using the end area method or Simpson's Rule, or by computer, with the new levels forming the input to the earthworks program.

The agent is usually satisfied with a computer print-out as the basis of the interim measure during the construction season, when earthworks quantities are out of date as soon as they are calculated. However, the volume for the final account is unlikely to be agreed on such a basis, unless the program has been developed primarily as a measurement tool rather than as part of a general engineering design package. Corrections will be needed in complex areas, and may be made either manually or by specially-written programs.

The best way of arriving at an agreed measurement is to carry it out jointly. The ICE Conditions of Contract (clause 56) require the contractor to 'attend' at the resident engineer's request to assist in the measurement and to provide any information necessary. The resident engineer must give reasonable notice but, if the agent does not co-operate, he can proceed on his own and his measurement cannot be questioned afterwards, except by an arbitrator.

The joint measurement can be conducted on a routine basis for those items which are permanently exposed, fencing, for instance, above-ground concrete and the general fixtures and finishings. Just before the monthly statement is drawn up an assessment can be made of the progress since the previous certificate. As long as there is agreement and the resident engineer is satisfied that the quantity is not grossly inaccurate, there is no point in collecting detailed dimensions as everything will remain visible for the duration of the contract and can be subjected to a more thorough measurement at a later stage. If there is disagreement on the quantity, or the particular item is the subject of a contractual dispute, then accurate and detailed measurements must be taken.

For items which are to be covered up or which, because of their location, will be difficult to measure in the completed works, the timing of the measurement must be dictated by the progress of construction. The staff concerned may have to make hurried arrangements to collect the information which they require, or rely on inspectors and foremen to make an agreed record as work proceeds. These items, for example, excavation of unsuitable material, filling into water or grouting, can only be measured once, so accuracy and agreement are essential if a settlement is to be reached.

Not only bill items are measured. Quantities associated with variations, unforeseen extras and claims must also be recorded to assist in determining any payments which may be due to the contractor. An agreed measurement should be sought even, indeed especially in cases where the resident engineer has rejected the principle of extra payment. The matter may not be closed and agreement on the facts, if not on liability, is of great help in speeding up the resolution of any dispute.

Most agents carry out a monthly reconciliation of materials delivered to the site against the quantities certified in the measurement. After making due allowance for materials which have been rejected, used on site but not measured (for example, in strengthening haul roads), used off site (in minor work done for adjacent landowners, for instance) and stockpiled without payment, the quantity unaccounted for should not exceed the normal wastage expected for that material. If there is any significant difference, an efficient agent will immediately seek to identify the loss, because the missing material means that his own costs cannot match those in the estimate.

He will check whether the measurement is short due to an error or to additional work being carried out on site without the issue of a proper variation order. He will also investigate the possibility of loss through unforeseen circumstances, such as excessive penetration of granular material into an unexpectedly soft sub-grade. It is most unlikely that any substantial discrepancy could escape this reconciliation for more than one month. Site staff who are aware of these procedures view with scepticism any late discovery of expensive losses alleged to be the liability of the employer. The existence of a series of agreed measurements, together with their supporting records, allows such assertions to be investigated objectively and makes spurious claims difficult to sustain.

Financial control

The financial control of construction projects is an important subject, and is rightly recognised as an area of specialist expertise. It is, however, an engineering expertise and one which site supervisors should study carefully. The resident engineer in particular carries a responsibility to the employer for the financial control of the contract at site level. He must be prepared to discharge that responsibility personally.

Otherwise he is restricting his authority and influence to technical matters, leaving the control of the economic aspects, which are just as important to the employer, to others.

Engineers have not responded to the challenges of financial management with uniform success. Brunel's memory was eclipsed by that of his father for almost a century after his death and he was judged a 'glorious failure' for the overspending and contractual disputes which plagued so much of his work. Locke, on the other hand, was respected by employers and contractors alike throughout his highly successful career, and long afterwards, for his close control of expenditure and his realistic grasp of the economics of construction. It is curious, and perhaps symptomatic of the problems of the modern profession, that Locke is relatively unknown in comparison with the younger Brunel, whose reputation has now attained almost heroic proportions.

The most important and effective financial control is that exercised at the planning stage. The funding and execution of a comprehensive site investigation and ground survey ensures that as much information as possible is available to the designer and the tenderers. Careful and thorough design minimises the need for variations and the incidence of unforeseen extra costs. Precision in the taking-off of quantities produces a bill which is a fair representation of the scope of the works, clarity in the preparation of the contract documents provides the tenderers with a complete picture of their obligations, and accuracy in estimating allows the employer to make proper financial provision for the cost of the scheme.

Once the contract price and the contractor's programme are known it is possible to produce a detailed monthly profile of expenditure. Typically, it will take the form of an S-curve. The peaks coincide with the mobilisation period at the start of the project, when large lump sums are usually due for the preliminaries, and with the summer construction seasons. The troughs correspond to the early months of slow preparation work, the winters and the closing stages when the time-consuming but usually low-priced finishing operations are all that remain.

Figure 12 shows the profile for a roadworks contract. The expenditure is shown in terms of the predicted monthly payments to the contractor and not as a running total. The start date is 1 February and the contract period is 24 months, but the contractor has programmed for a finish before the second Christmas, a duration of 22 months.

The first interim certificate is estimated to be a high one because although many of the preliminaries spread over the duration of the contract, there are a number of lump sums (for establishing the site offices and the laboratory, for instance), which will require early settlement and have been 'front-end loaded'. Thereafter payments run at a lower level, picking up after the Easter break when earthworks are expected to be in full swing and the initial low-earning period on the structural work, when most effort is directed into temporary works and excavation, will be replaced by more lucrative operations such as piling and concrete foundations.

The summer months bring a pronounced peak when it is assumed that the contractor will be running earthworks and carriageway construction (up to road base level) simultaneously, and also pressing on with the structures.

No payment is forecast for December, as the small amount of work programmed before the Christmas shut-down is unlikely to bring the valuation above the minimum level for a certificate. After the bad weather of January and February, the contractor should be able to make a faster start than in the first season. With most of the earthworks complete by the end of the Spring, carriageway construction will continue through the second summer with the high-value surfacing items going ahead during July to September to produce the peak spending period of the contract.

The rapid decline in certified payments after the end of the surfacing marks the shift of effort to low-rated items such as verge finishings, soiling and seeding and white lining and to general remedial works, which generate no payments.

The profile assumes substantial completion to be achieved as programmed and allows for the release of half the retention money at that date.

As the works progress, the resident engineer compares actual expenditure against the profile, amending it as required and warning the employer of any significant changes so that the necessary adjustments to the budget can be made. Any major revision of the contractor's programme would require a new profile.

The interim payment certificates and the cost profile are useful to the employer, but can hardly be said to keep him fully informed of the progress of his project and the way his money is being spent. A monthly report in narrative form accompanying the certificate provides a link between cost and construction.

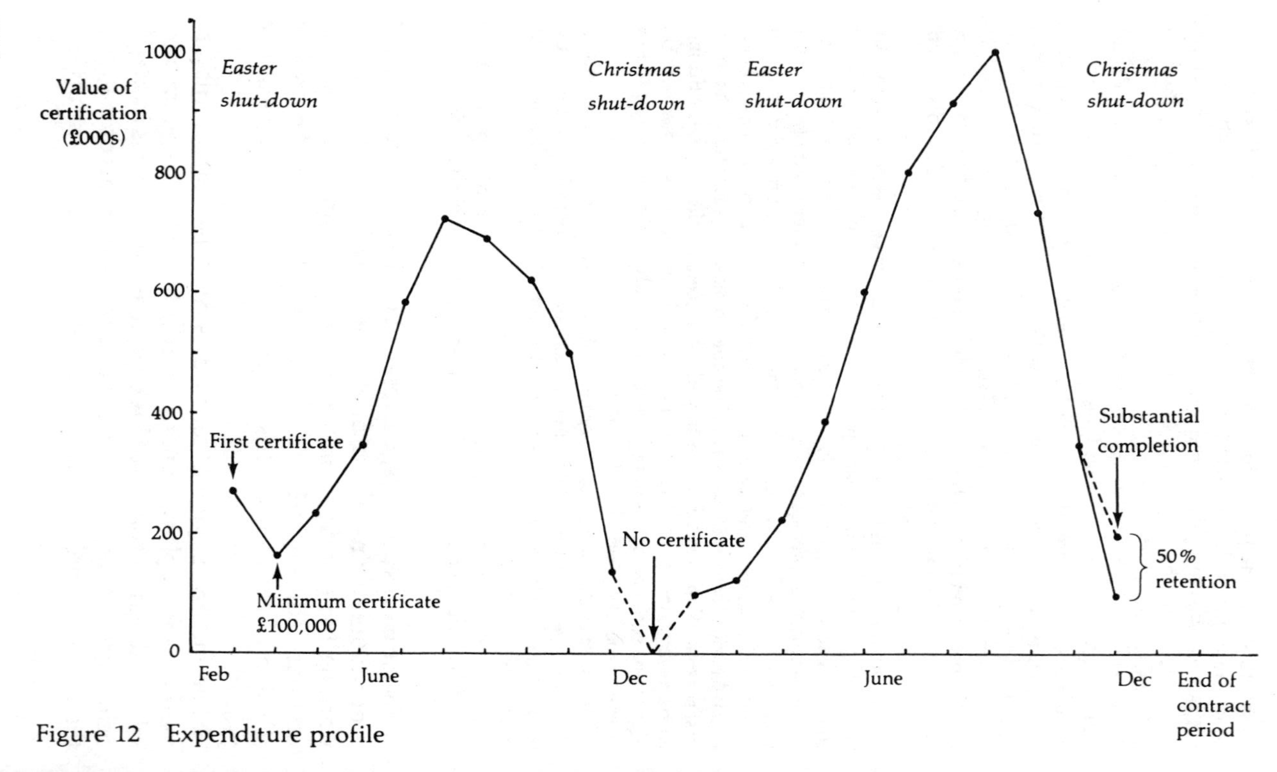

Figure 12 Expenditure profile

The report includes brief notes on the progress of each major part of the works, a description of any other work being undertaken, by statutory undertakers, for instance, and a general summary covering the current assessment of progress against the programme, the latest position on disputed items and any other matters the resident engineer considers relevant. As the interim certificate does not describe the finances of the contract in the most understandable way, nor cover all the costs, committed and projected, of which the employer should be aware, a financial statement, such as that shown in Figure 13, is a useful addition to the monthly report.

The statement in Figure 13 corresponds to the interim certificate shown as Figure 11. The bill figures are presented differently. The roadworks is not sub-divided into main carriageway and side roads but into the two major operations of drainage and pavement construction. There is also less detail. The value of each main activity is expressed as a percentage of the appropriate total in the bill, giving an indication of the progress of the works, albeit an approximate one, which is not provided in the certificate.

The resident engineer's estimate of the final contract price is given, based on his current assessment of the value of variations and extras. The full total of extras claimed by the contractor to date is also included. Any particularly expensive variations would be listed separately.

The value of other work outside the main contract, in this case the diversion of gas and electricity services and British Rail's charges for flagmen, lookouts and supervisors, is provided, allowing the employer to assess the latest position on the cost of the whole project.

Additional costs due to variations or extras must be closely monitored and kept to the minimum necessary for the satisfactory completion of the works and the proper discharge of the contract. The documentation in support of any claims must be submitted promptly and in a format which permits checking and verification. Despite the opportunity for future revision of any interim certificate, the site staff should insist wherever possible on full and accurate particulars rather than 'round figures' and scrutinise the agent's submission with a view to making a firm recommendation for settlement rather than an 'on-account' estimate.

When variations and extras arise, the employer expects the resident engineer to provide an accurate assessment of their financial effect, moreover, he will not expect any interim

NAME OF CONTRACT

FINANCIAL STATEMENT

September 1984

(A) Measurement

	£		£
General Preliminaries			319,924 (42% of value)
Roadworks:			
- General (Fencing, Earthworks)	444,188		
- Carriageway (Drainage)	209,651		
- Carriageway (Pavement etc.)	57,442		711,281 (35%)
Structures:			
- Railway Bridge	137,938		
- Footbridge	65,100		
- Others	78,108		281,146 (38%)
Dayworks:			6,000
Balancing Item		PLUS	16,688
	TOTAL		1,335,089 (33%)

Certified Value in Interim Certificate No. 7 (after allowance for Extras, Materials on Site and Vested, Contract Price Fluctuation and Retention) £1,373,884

Payment due on Certificate No. 7 £249,776 (passed 22.10.82)

(B) Current Estimate of Additional Costs

(i) Ordered Variations to 30.9.82 -	additional costs	£105,000
	less savings	£21,000
		£ 84,000
(ii) Extras and claims paid "on account" N.B. value notified: £123,000		£ 4,000
		£ 88,000

Current Estimate of Contract Price (base dated) £4,088,000

(C) Supplementary Costs

Gas Board	£ 47,000
Electricity Board	£161,300
British Rail	£ 12,400
	£220,700

(No increases over estimated costs have been identified to date)

Figure 13 Financial statement (see Figure 11 for corresponding certificate)

settlement to be significantly exceeded in the final account. Reliable cost prediction and control are essential to the employer's management of the contract, but the site staff may misinterpret this legitimate concern with expense (particularly additional expense) as pressure to resist all claims and 'screw the contractor down'. Regrettably, there are cases when the pressure is not imaginary. Resident engineers, their staff and employers must never forget that controlling costs does not mean witholding payments to which the contractor is properly entitled.

Final account

After the works have been completed and the maintenance period has expired with the satisfactory execution of all remedial and outstanding work, the Engineer (not the resident engineer, for this power cannot be delegated) issues the maintenance certificate. This act moves the contract into its last phase: the settlement of the final account between the employer and the contractor.

The procedure for preparing the final account is set out in clause 60(3) of the ICE Conditions of Contract, and the importance of this phase of contract administration is recognised in the reservation of all powers and authority to the Engineer himself. If the resident engineer and his team have kept the measurement process moving forward during the construction stage, so that the interim certificates represent a fair evaluation of all the work (billed items, variations and extras), the settlement of the final account should be easy. This ideal situation depends not only on the availability of resources on site and the pace of the works but also on the agent's level of commitment and co-operation. Usually some combination of these factors hampers the production of a complete final account concurrently with the progress of the works. It is, nevertheless, a target worth aiming for and the final settlement of any part of the bill, however small, is a useful gain.

The contractor is allowed three months after the issue of the maintenance certificate to present a statement listing in full, and with all the necessary supporting details, the value of the measured work and any other sums 'which he considers to be due to him under the contract', that is, claims. In fact, most of the documentation should already be available to the Engineer: the contractor's valuation of the measured work has been submitted in previous monthly statements; the records and other details of costs covering the ordered variations have been presented during the consultation with the resident engineer; and 'full and detailed particulars' of all claims have been delivered to the site staff as early as possible, as required by the ICE Conditions of Contract. There is no reason why, in this instance, the theory of the contract should not be borne out in practice. The final statement may well include certain refinements of detail. A few remaining approximations may be replaced by accurate figures, some errors may be corrected and, in the case of those claims which continued their effect through to the end of the construction stage and perhaps into the maintenance period, it may be that the complete financial picture is only now available. Nevertheless, after the passage of almost 15 months (assuming a maintenance period of one year) since the completion of the works, and in view of the contractor's obligation to pass on all information relevant to the measurement or claims as soon as practicable, the appearance of anything new in the final statement would be surprising and difficult to explain.

Although the measurement of some large or complex bill items may be resolved only at final account stage, it is usually the settlement of varied work and claims which absorb most effort. Consultation between the resident engineer and the agent results in many variations being agreed as the interim measurements proceed, and these are incorporated into the final account. A practical, co-operative approach can solve many problems, but it is inevitable that some rates and prices cannot be settled by agreement and must be determined or fixed by the resident engineer. The results of this exercise are considered by the Engineer for the works along with the contractor's final statement.

The process of fixing rates or prices follows the basic method of estimating which can be expressed by the formula:

$$\text{rate} = M\left(1+\frac{m}{100}\right) \times L\left(1+\frac{l}{100}\right) + P\left(1+\frac{p}{100}\right) + X$$

where M = material cost per unit of quantity
m = percentage on-cost on material
L = labour cost per unit of quantity
l = percentage on-cost on labour
P = plant cost per unit of quantity
p = percentage on-cost on plant
X = any other miscellaneous cost

The operation is broken down into its constituent parts and the input from the three major elements, materials, labour and plant, is assessed from experience, supplemented as necessary by estimating handbooks and trade literature. The cost is calculated using the same sources together with schedules such as those published by the Civil Engineering Construction Conciliation Board or the Federation of Civil Engineering Contractors, details of local hire rates and quotes from suppliers. The percentage on-costs may be obtained from the agent, but he is entitled to refuse such a request and leave the resident engineer to make his own assessment, which must be based on previous experience, figures obtained from other contracts and a careful study of the bill rates. The on-cost covers the agent's site offices, compound, staff and general establishment costs, temporary works, saftey measures and supplies of fuel and water, etc. It also carries a proportion of head office overheads and various insurance and financing charges and last but not least it includes, subject to adjustment by the balancing item if there is one, the contractor's profit margin.

Rate fixing is a complex and difficult task, best undertaken in co-operation with the agent. In the absence of such assistance the resident engineer must make a unilateral evaluation, the detailed calculations of which he would be most unwise to release to the agent (as discussed in Chapter 10), although the Engineer does, of course, have them to hand when he assesses the contractor's final statement. The final certificate contains the engineer's confirmation or revision of the rates fixed by the resident engineer. Thereafter the contractor must either accept them or pursue the matter as a contractual dispute.

The Engineer also has before him as he prepares the final certificate the contractor's claims, with their supporting particulars and the results of the resident engineer's investigations. Extra costs which have already been agreed on site are confirmed, and outstanding claims are considered and evaluated. Those claims rejected in the final certificate must either be pursued as disputes or withdrawn, for it is only in exceptional circumstances, such as the discovery of new evidence, that consultation outside the disputes procedure can be re-opened at this stage.

The Engineer has three months in which to prepare and issue his final certificate, although this period can be extended if the contractor has submitted insufficient information with the statement. The certificate states the amount payable to produce, in his opinion, final settlement of the contract. Usually this is a sum due to the contractor but, in cases where interim certificates have been valued at too high a level resulting in over-payment, or when liquidated damages are due, it may be the employer who receives the final balance. In either case, the transaction must be completed within 28 days of the issue of the final certificate.

Neither the maintenance certificate nor the final certificate concludes the contract. For six years (12 in the case of contracts under seal) the contractor remains liable to the employer for defective work or materials and the employer continues to be liable to the contractor for claims arising out of the contract, although in the latter case the question of proper notice may raise insuperable problems unless the claim was submitted before the final certificate. Similarly, the Engineer's powers are not brought to an end, for he may be recalled to give his decision on a contractual dispute arising out of the dissatisfaction of either party with his final certificate or developing within the six — or twelve — year limitation period.

Chapter 12
Record keeping

It is said that in the vital tasks of assessing payments, determining liability and settling disputes there are only two essential requirements: records, and more records. Without a detailed record of the progress of every aspect of the project, the Engineer for the works will be seriously handicapped in drawing up the final certificate and adjudicating in disputes between the parties to the contract. If the information necessary to make a reasoned engineering judgement is lacking, there is a strong probability that the decisions of the Engineer will be subjected to criticism and review by others and that the interests of the employer in particular and the contract in general will suffer. Volume of records, however, is not enough. They must also be both contemporary and verifiable.

The site records

There can never be a comprehensive, universal list of all the records the resident engineer should keep, as on each contract the nature of the works, the amount of delegation and the arrangement of the channels of communication generate their own requirements. It is possible, however, to identify in general terms the types of records which are fundamental to effective supervision, and which provide a framework around which a working system can be built, but, to use a phrase of great value in site correspondence, 'this list is not exhaustive':

(1) all correspondence between the resident engineer and the agent, including variation orders, CVIs, approval forms, etc.;
(2) all correspondence between the resident engineer and the Engineer, the employer and 'third parties';
(3) the minutes or notes of every formal meeting;
(4) plant and labour returns, as submitted and as corrected;
(5) measurement records, including dimension books, timesheets, delivery notes etc.;
(6) daywork records, as submitted and as corrected;
(7) interim statements, as submitted and as corrected, with copies of all supporting particulars, and interim certificates;
(8) level and survey books, covering both checks on setting out and completed work;
(9) laboratory results and any other testing data;
(10) weather records;
(11) progress photographs;
(12) construction drawings;
(13) administrative records such as leave and sickness returns, accident reports etc;
(14) site diaries.

Correspondence files, together with site diaries, form the heart of any record system. The original of every incoming letter and a clear copy of every outgoing, with any enclosures, are placed on file under the appropriate subject heading. Each site team will develop its own filing system, but a good starting point is to set up a file for each section of the bill and the specification, and for such essential subjects as land ownership, statutory undertakers, site safety, programme, setting-out, sub-contractors, suppliers and public relations. The main requirement is ease of reference, and the existence of general 'dustbin' files, such as 'Instructions to the agent', lead to confusion as their contents are so numerous and varied that it is difficult to find any particular topic and harder still to follow it through. Wherever there is doubt about the most appropriate file heading, or when letters deal with more than one subject, extra copies should be made and cross-filed. To ensure the files are complete all incoming and outgoing correspondence must be logged in a register and given file references. The register should be checked against the files at frequent intervals to ensure that nothing goes astray. A second aid to making the system as comprehensive as possible is a 'day-file' in which a copy of each item of corrrespondence is filed by date rather than by subject.

All the records submitted by the agent for checking, such as plant and labour returns, dayworks sheets, measurement particulars, must be carefully preserved. Thus a working copy is used for the check and any consequent revisions and a file copy is kept 'clean'. Both are preserved.

Laboratory records and other test results are usually entered on standard forms and filed by subject. When the contractor is being issued with some, but not all, results it is prudent to indicate on each sheet whether or not it has been copied to the agent. The presentation of such a mass of information in a form understandable and useable by the supervisory staff requires more than simply efficient filing. The use of graphs to plot the distribution of results or to indicate trends by cumulative sum ('cusum') techniques is an effective way of handling figures in bulk. The establishment of the link between quality control and end product requires a different form of visual representation. Schematic drawings of parts of the works — for example, the earthworks and the different structures or structural components—can be specially prepared to display the location of all tests or samples and their corresponding results (when the nature of the result does not permit clear presentation, the date can be quoted to assist easy reference to the files). Other relevant details, such as the extent of any material or work which has been rejected and removed, can also be incorporated. Figure 14 shows a section of the schematic drawing covering pavement construction on a new dual carriageway.

A photographic record of the progress of the contract can be of great assistance in settling disputes over the timing of particular operations and the condition of various parts of the works which are no longer visible. A series of photographs

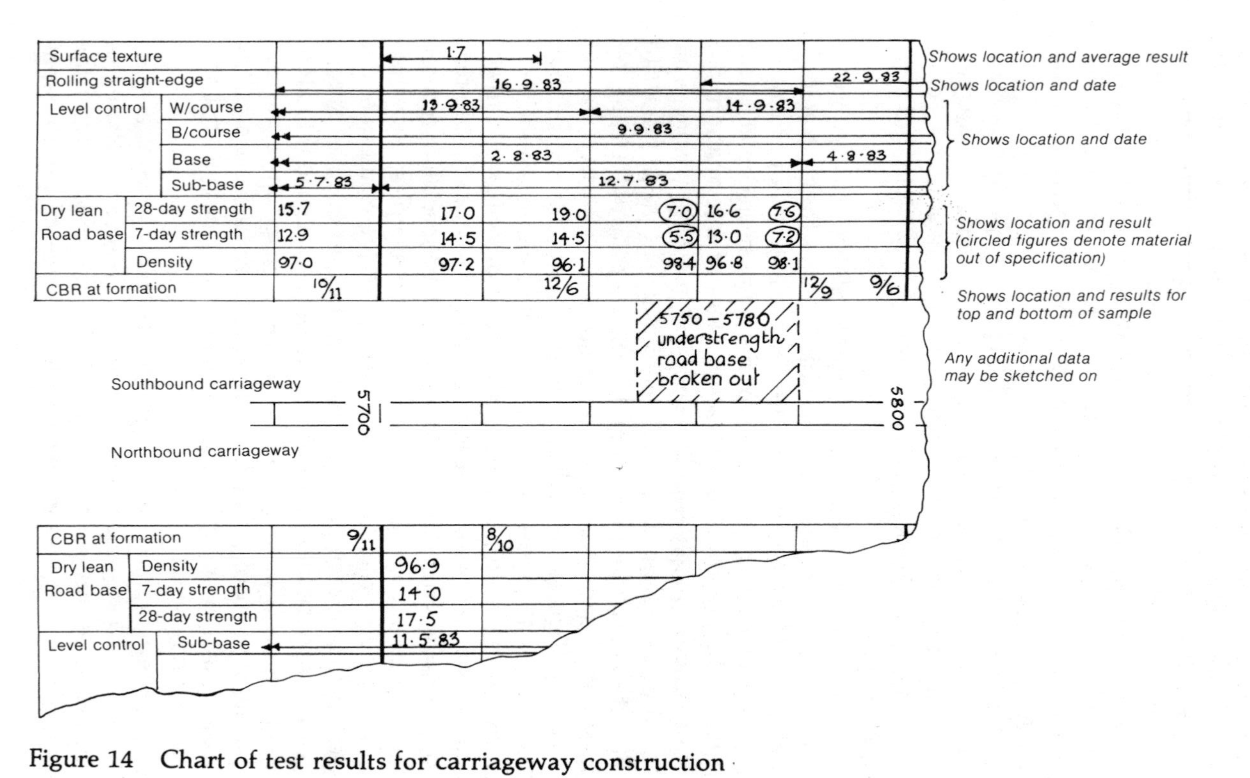

Figure 14 Chart of test results for carriageway construction

taken from the same position at regular intervals, say once a month, provides coverage of the main features, supplemented if required by individual photographs recording particular subjects, such as the honeycombing on a rejected section of concrete retaining wall or the extent of flooding caused by a period of exceptional rainfall. Commercial photographers have the equipment and expertise to produce results with the clarity and detail necessary for evidential purposes, but they cannot be standing by to provide an instant service, so a site camera must be available to maintain a continuous record. Every photograph must be accompanied by details of the date, the subject and the position and direction from which it was taken.

As the work proceeds a set of general arrangement drawings should be revised to show all the changes resulting from the contractor's proposals, adoption of permitted alternatives, different topographic or sub-soil conditions and variations. These construction drawings serve two purposes: they prevent errors arising from staff consulting the original works drawings and consequently relying on information which has been superseded and they form the raw material from which the final 'as-built' drawings are prepared for the employer.

The importance of the site diaries cannot be overstated. They provide a complete narrative of the progress of the works and the activities of the resident engineer and his team. Everyone must keep a diary, not only to ensure that the maximum amount of detailed information is collected but also to permit cross-checking to confirm the truth or otherwise of disputed statements. A standard type of diary must be used, preferably of the self-carbonating duplicate variety, batch-printed with each page carrying a unique number. There should be no printed dates, allowing each entry to be unrestricted in length. Writers must never leave a gap at the end of one day and start a new page for the next; they must draw a line to conclude the entry and begin again with the new date immediately below so that it is demonstrably clear that the insertion of further material at a later stage is impossible. The object of these precautions is to establish the diaries as a routine, contemporary system of records which will be acceptable as reliable evidence in any subsequent proceedings.

The diary must provide a factual record of events on site, conversations with the agent's staff and others, instructions issued and weather conditions. The entries must be accompanied by full details of the time, location and personnel involved. It is acceptable to include statements of opinion, as long as these are clearly identified as such. In contrast to the approach recommended for site correspondence, it is better to write too much than too little. Regular inspection of the diaries by the resident engineer ensures that they are up to standard, but he must remember that his diary is the most important of all and deserves considerable time and effort.

The length of a diary entry depends not only on the level of activity on a particular day but also on the writer's style. Note form is satisfactory provided it can be understood by any reader, but over-enthusiastic use of abbreviations is not helpful, nor is the adoption of some form of 'personal code'. Consequently, engineers and technicians must inevitably spend a significant part of their day collecting information for their diaries and writing it up. Rough notebooks are very useful for jotting down basic details and reminders, but even better are pocket tape recorders which will pay for themselves many times over in the efficient use of staff time. It is essential that every team member disciplines himself to complete his diary at the end of each day. Exceptionally, a rough draft can be made out and transferred into the diary the next day but this practice is not to be encouraged, for it undermines the principle of keeping the site diaries as a contemporary record. Writing up the diary (Figure 15) at the end of the day must become a habit.

Few engineers or technicians can say they enjoy filling in their diaries, but the inspector has not been born who views this task with anything other than thinly-veiled disgust. Nevertheless their contributions are of great value because they spend almost all their working day on site in close proximity to the contractor's operations. They, of all the site team, are best placed to record the detailed deployment of plant and labour, the movement of materials and the progress

Royal County of Berkshire – Department of the County Surveyor – **Site Diary**

Project – A34 SHAW BYPASS	Name – J. J. SMITH	No. 3401

08.15 Cold and clear but no overnight frost.
Ground still too wet for earthworks.
Drove to the foundation excavation at the north bank of the river and met K. Harris (sub-agent) and J. Thomas (materials engineer). Inspected the exposed bottom and agreed that an area 2.5 × 3.0 m in the south-east corner was soft and should be removed and replaced with crushed stone. JCB 805 on site and available to remove this soft spot. Ordered K. Harris to proceed — formal variation order to follow. Arranged for an inspector (N. Durham) to attend and measure the depth of excavation to a sound bottom.

08.50 Returned to site office via subways (no work) and bridge 5. Piling rig still standing on access road. Ganger says this is due to clutch failure in the tractor unit. I think the contractor may claim that the rig was held up by the Electricity Board's work but this is not so; the trench could be temporarily backfilled where it crosses the access road if the rig was ready to move.

15.00 Prolonged showers. Saturated ground is causing rapid run-off into open excavations. Toured section with P. Hempstead and N. Durham — flooding already apparent at subways 2 and 3 and bridge 5; cut-off ditch working OK at subway 1 (2 steelfixers now at work at west end), pump in action at bridge 4 (river) where blinding has been completed. Took record photos of flooded excavations. Spoke to K. Harris who says that a pump is coming to subway 2 tomorrow so that RB22 can be moved down from the compound to restart excavation. Warned him that bottom may be saturated and not acceptable for blinding. Passed access track to bridge 5 – rig still standing.

16.45 CVI in from contractor re. soft spot at river bridge — same as the VO issued this morning except it says payment will be at daywork rates. Phoned K. Harris and told him this was not agreed or even discussed — there is a bill rate. He agreed it was an error and would withdraw the CVI as our VO had now arrived.
17.15 Checked barriers and lights at subways 1 and 2 — OK.
Prepared last week's progress report.
18.00 Off site — raining steadily.

J. J. Smith

Figure 15 Extracts from a site diary

and problems of the activities they are supervising. The most efficient way of getting this information on to the record is to issue inspectors with daily diary sheets, divided up into headed columns for time, location, activity, details of plant and labour and general remarks. Faced with limited time and difficult working conditions, the inspectors can make quick and frequent entries on these sheets with minimum interruption to their work. Taken together they should account for the whereabouts and activity of all the plant and labour on the site.

The assistant resident engineers, or perhaps the section engineers, use the site diaries to draw up a summary of the main operations and events for each week and month of the contract. This provides a general record of progress for ease of reference, eliminating the need to search through a set of site diaries to find the date of a particular operation. It is, however, only a straightforward chronological statement of the work which is in progress, or has been completed, during the period under review.

So important are the diaries as the principal record of the contract that it would be most irresponsible to risk their loss by keeping both the originals and the carbon-copies together on the site. The carbons are best collected and bound together each week for despatch to head office for safe keeping. All completed diary books and the bound daily sheets prepared by the inspectors should be kept in a locked, fire-resistant cabinet. Diary books which are in use should never be left lying around the office.

The problem of 'contemporary records'

Whenever the contractor seeks to claim extra payment, he is bound to provide supporting evidence detailing not only why the employer is considered to be liable for these costs, but also how they have been incurred. Clause 52(4) of the ICE Conditions of Contract sets out the requirements for the evaluation of a claim, and foremost among them are the contractor's 'contemporary records'.

There is no explanation of what form these records must

take, nor of what information they must include, except to say that they have to provide 'full and detailed particulars' of the value and grounds of the claim. Clause 52(4) is not noted for the clarity of its drafting. If a definition is to be built up from first principles, the meaning of 'contemporary' appears to present no problems: records kept at the time of the events to which they refer. However, there is the difficulty that neither commonsense nor the contract would expect all such records to be submitted to the resident engineer immediately. In many cases it is only after the events giving rise to the claim have occurred that the agent realises the situation and puts together his evidence. Thus, although the records may have been taken at the same time as the events which they cover, there may be some delay before they come before the resident engineer in support of a claim. It is not enough for them to be merely 'contemporary', their content must also be verifiable. This requires the agent to show that his records are, like the site diaries of the resident engineer's staff, part of a routine system which is constantly kept up-to-date, and that they conform with other documentation such as the weekly plant and labour returns. The longer the delivery of records is delayed, the more complete and comprehensively cross-referenced they must be to permit verification.

Similarly, there is no definition of what constitutes proper 'records', although the ICE Conditions of Contract do offer the very subjective advice that they should comprise what 'may reasonably be necessary to support any claim' the contractor may wish to make. The nature of the claim will govern the content of the supporting evidence, but as a minimum it is necessary to provide timesheets for plant and labour, delivery notes and agreed dimensions for materials and invoices or accounts to substantiate costs. It is also essential to prepare and present to the resident engineer a reasoned argument, not necessarily final in its content or conclusions, which demonstrates the link between the records and the contractual justification for the claim.

Contractors and their agents sometimes assert that keeping the contemporary records and 'full particulars' required by resident engineers is made difficult by their natural inclination to 'get on with the job' and their understandable reluctance to transfer limited resources from that effort to the special task of collecting the information for these records. This is a spurious argument for two very good reasons. Firstly, there is the contractor's need to survive in a commercial environment, which necessitates that one of the agent's principle responsibilities is to account for the cost of all his activities. The site is designated a 'cost-centre' and is charged for the wages of the labour force, payments to sub-contractors and suppliers, and the costs of materials and hired plant. When materials and plant are supplied from within the contractor's own organisation, charges are nevertheless made and to ensure their accuracy the agent will keep details of quantities, wastage, hours worked, standing time and other relevant information. Sub-contractors may obtain some or all of their supplies and equipment from the agent, and have the costs deducted from their account for the sub-let work. These contra-charges must be substantiated by itemised invoices. Consequently, a formidable amount of paperwork is involved in keeping, presenting and settling the accounts of a construction site, all of which must be regularly and routinely processed, subjected to checks for accuracy and linked to the bill items under which the contractor is paid. A better source of 'contemporary records' is difficult to imagine.

Secondly, there is a sound economic reason for providing the necessary resources to collect the information: no records mean no claim and no payment. If the agent is making a substantial claim which he considers to be fully justified, then it must make good financial sense to invest in the staff time needed to support it, either by working overtime or by bringing in extra personnel. Temporary staff can be obtained for short assignments through the employment agencies which are active in the industry, and the agent should be able to balance this relatively small cost against the substantial sums he expects to recover.

Quantity surveyors and auditors

Employers have always been concerned over value for money

in their civil engineering projects, but the economic climate of the modern world and the importance now given to accountability in the public sector have combined to make it essential that contracts are not only closely controlled financially, but are seen to be controlled.

Engineers, partly due to their training, partly by inclination, have tended to give greater priority to the design process than to financial management. Furthermore, and perhaps more damaging to their professional reputation, engineers have given the impression that they find problems of design much more 'interesting' than details of cost. Technical papers and articles rarely include information on prices or rates, and many site staff have little or no idea of the cost of the items in the bill.

Employers, like Nature, abhor a vacuum, and quantity surveyors have extended their influence in recent years from the building sphere into civils work. Quantity surveyors are prepared, if some engineers are not, to involve themselves in the intricacies of hard decisions of measurement, estimating and financial control. This willingness has been repaid by the confidence of promoters and employers and many engineers are concerned that the independence and supremacy of their supervisory role in the contract is being seriously undermined.

There is a well-known convention on site that the phrase 'with respect' is used to precede a statement likely (or even calculated) to give offence. It must, however, be said, *with respect*, that whilst the quantity surveyor can play a useful and valuable part in the site team, he is not essential to the proper administration of the contract. A fully trained engineer should be capable of carrying out all the functions in the quantity surveyor's specialisation, as well as displaying a broader appreciation of the practical and theoretical aspects of construction.

Thus, on a contract executed under the ICE Conditions of Contract, there is no formal office of quantity surveyor to compare with that found in the JCT form. The Engineer for the works, or his representative where appropriate, has the sole responsibility for measurement, determining valuations, settling liability and assessing claims, subject only to both parties' right to seek redress through arbitration. Advice may be obtained from wherever it is reasonable to seek it, and quantity surveyors are a good source of information and informed opinion, but the Engineer must ultimately act for himself. The proper role of the quantity surveyor, or the measurement engineer, for that matter, is in relation to quantum (the amount) and not principle.

Auditors seem to have acquired an odious reputation which can hardly be truly deserved, their role has been described as 'coming in after the battle to bayonet the wounded'. Their real duty, which may be established by law or by the regulations of the organisation by whom they are appointed, is to ensure that the accounts represent a 'true and fair' record of the transactions which they purport to cover. The auditor is concerned with financial accuracy and propriety but he is not required to involve himself positively with matters of management or professional judgement (although, of course, he will report on the financial effects of what he may consider to be bad management or judgement). A judicial description of long-standing refers to him as a 'watchdog but not a bloodhound'.

The auditor, whatever his legal duties and status in the organisation he serves, is not mentioned in the contract and thus has no standing to intervene in the supervisory process. The ICE Conditions of Contract contain no qualifying wording and so the Engineer's certificates, interim and final, are not paid 'subject to audit' and may be re-opened, as far as the contractor is concerned, only by an arbitrator.

The all-important principle is that the Engineer is granted certain independent powers under the contract which he alone may exercise, subject to limited rights of delegation. These powers cover the whole range of measurement, payment, assessment of liability and evaluation of extra costs and delays. The contractor is entitled to expect the Engineer to discharge these duties unfettered, and can look upon any attempt by the employer to interfere or to operate secondary systems of control as a breach of contract. The Engineer may take advice, but not instructions. He may inform the employer of problems of extra costs and give his views on how they may be mitigated, but he cannot have his powers to certify payment restricted or his certificates amended. The employer can express his dissatisfaction with the actions of the Engineer through the disputes procedure or he may seek redress in the courts if he feels his professional adviser has been negligent, what he cannot do is reverse the decisions of the Engineer or substitute those of his other advisers in their place.

Chapter 13
Claims and disputes

Much of the folklore of contract administration, and the majority of the literature on that subject, is focused on claims and disputes. No observer of the construction industry can fail to note the very significant amount of intellectual effort, and expense, which is put, not into solving potential problems of contract management, but into determining liability for the consequences of those problems.

Despite this concentration on claims, many contracts are completed without significant disputes of any kind. Not all contractors are 'claims merchants' and not every claim is made on spurious grounds. Nevertheless, the probability of a serious dispute arising on a contract is sufficiently high, and the proper handling of the matter so important, that the site staff cannot afford to ignore the subject nor to learn their lessons as events develop.

Grounds for a claim

To establish a claim it is necessary to state the grounds on which it can be made. There are two ways of approaching this problem of definition. The agent generally identifies a situation on the site which is causing, or seems likely to cause, extra costs and searches in the ICE Conditions of Contract for the appropriate clause to cover his claim for reimbursement. The resident engineer, on the other hand, usually adopts a more legalistic approach. He is concerned to recognise the presence of any employer's liability and, where a principle for reimbursement has been established, to assess what extra costs or delays are actually attributable to it. In spite of the different approaches, the solution is the same: claims must relate to clauses, since payment must be related to the terms of the contract.

On contracts governed by the ICE Conditions of Contract, the following clauses represent the most common grounds for claims:

Clause 5	documents not mutually explanatory;
Clause 7	late issue of drawings or instructions;
Clause 12	unforeseen physical conditions or artificial obstructions;
Clause 13	extra costs or delays caused by instructions to work to the 'satisfaction' of the Engineer;
Clause 14	delayed approval of, or modifications to, the contractor's proposed methods of working;
Clause 20	the 'excepted risks';
Clause 31	activities of the employer's own workmen or other contractors;
Clause 36	unforeseen or unspecified tests;
Clause 38	unnecessary exposure of work executed in accordance with the contract;
Clause 40	suspension of the works;
Clause 42	late possession of the site;
Clause 44	extension of time;
Clause 49	work arising during the maintenance period which is not covered by the contractor's obligations;
Clause 52	valuation of variations, dayworks and rate fixing;
Clause 56	changes in rates due to changes in billed quantities;
Clause 59A	directions to enter into a nominated sub-contract to which the contractor objects;

Clause 59B	forfeiture of a nominated sub-contractor;
Clause 60	incomplete or overdue payment.

A claim is generally taken by the resident engineer to mean a request for payment arising out of some occurrence not envisaged in the contract, or for which the contract machinery has not produced an agreed settlement. Thus the agent's submissions in respect of variations and dayworks (clause 52) and changes in rates due to changes in quantities (clause 56) would not be considered as claims by the site staff unless and until the process of consultation had run its course and the contractor had expressed himself dissatisfied with the result. Similarly, a request for an extension (clause 44) would be viewed as a claim only if the contractor disputed the resident engineer's assessment.

As far as the supervisory staff are concerned, claims are made under the terms of the contract and in relation to specific provisions—the Engineer has no authority to deal with any other form of action. A contractor can, however, bring a claim against an employer under the general law of contract or in the branch of the law known as tort, which embraces such civil wrongs as misrepresentation and negligence. The remedy would be damages and would cover not only the extra costs of additional work, disruption or delay, but also loss of profit. The action would be brought in the courts and the Engineer and his staff would play no part, except perhaps as witnesses.

Claims founded on neither the terms of the contract nor general legal principles may, in exceptional circumstances, be entertained by the employer although he will be under no obligation to settle. Such claims may arise out of an event, not contemplated by either party and so not an implied term, which causes significant loss to the contractor although not preventing him from executing the works.

The Employer, having received the benefit of the completed works, may wish to make an 'ex gratia' payment to mitigate the contractor's losses. Only the employer can authorise such a payment — the Engineer may give advice, but has no power to certify — and then only if there is a good explanation to satisfy the auditors, shareholders or elected representatives who are likely to question action of this kind very closely. An example of the circumstances which can result in 'ex gratia' payments is the oil crisis of 1973, when the rapid and unexpected rise in the cost of oil and oil-based products caused severe losses to contractors working under the fixed price contracts then generally in use. Several major employers, including the Department of Transport, arranged 'ex-gratia hydrocarbon payments' to contractors who could demonstrate, from an inspection of their accounts, that they had suffered losses directly attributable to the rise in oil prices.

Notice and submission

The terms of the contract define how claims are notified and submitted. Clause 52(4) of the ICE Conditions of Contract, although rather confusingly combined with the procedure for valuing variations and couched in unnecessarily obscure and impressive language, sets out the general requirements for *all* claims:

(1) Notice must be given in writing to the Engineer (or his representative) if the contractor intends to make a claim.
(2) The contractor must give notice 'as soon as is reasonably possible after the happening of the events giving rise to the claim' (although in respect of claims against valuations of variations and changes in rates due to changes in quantities the time limit is set more precisely at 28 days).
(3) As soon as he is aware of a possible claim, and without any instruction from the supervisory staff, the contractor must keep contemporary records.
(4) The Engineer (or his representative) can instruct the contractor to keep particular records once he has received notice of a claim, but this does not alter the contractor's absolute obligation to keep his own records.
(5) 'as soon as is reasonable' after giving notice, the contractor must submit 'full and detailed particulars', not only of the amount claimed but also of the grounds for

the claim, and continue to present further accounts as appropriate.

(6) If the contractor fails to submit proper records and particulars, to the extent that any investigation of his claim is prevented or 'substantially prejudiced', he may lose all or part of his entitlement to reimbursement.

There can be no standard description of the form a claim submission should take, for each situation has its own special features and must be treated on its merits. Nevertheless, every submission has to perform three fundamental tasks:

(1) It must identify the principle on which the claim is founded, by reference to the ICE Conditions of Contract.
(2) It must demonstrate the chain of cause and effect by which the grounds of the claim are linked to the extra costs and delays the contractor is said to have suffered.
(3) It must enumerate and substantiate all extra costs.

A submission will therefore include an explanatory statement, a narrative and full supporting documentation, all of which must link together to answer the questions 'how?' and 'why?' as well as 'how much?'. It is not enough for the contractor to indicate a situation which might give rise to a claim, a change in ground conditions, for example, and then to set out his total earthworks costs, subtract the corresponding payment at bill rates and ask for the difference. The submission must show how the earthmoving operations had to be changed, why this was not foreseeable at tender stage and how much the changes added to the original estimated cost.

The question of notice is always difficult. Whereas many claims arise out of a specific event, such as the late issue of a drawing or failure to provide access to a part of the site, there are always 'grey areas' which involve matters such as the resident engineer's instructions to do work to his 'satisfaction', or the nature of the sub-soil in an excavation. These 'grey areas' are not immediately apparent as causes of extra costs or delays. However, it is unlikely that a significant loss will go unnoticed for more than one interim statement and thus there must be a very good reason for a notice being delayed by more than two months. Having given notice, the agent must keep records, but he is not given a strict schedule, the only requirement being to submit particulars 'as soon as is reasonable'. There are some claims for which a submission might reasonably be expected to follow notification almost at once; for instance, when the contractor is instructed to carry out an unforeseen test on a component. There are other situations, however, in which the collection of evidence and the assessment of the full effects is bound to take time. Claims in relation to ground conditions are typical examples of this category. There can be no hard-and-fast rule for such cases, except to stress that, in the absence of complete information, interim partial submissions may be acceptable and a most useful aid to verification of detail. If a contractor has a good case, it cannot be harmed by the presentation of particulars as they become available. Reluctance to provide the site staff with evidence, particularly when it relates to work which should be subject to contemporary examination, deserves to be viewed with great suspicion.

A case frequently cited as authority on the timing of notices and submissions is Tersons *v* Stevenage Development Corporation (1965) 1 Q.B.37. The action was settled, in favour of the contractor, in the Court of Appeal in 1965 but the claim arose out of two sewerage contracts undertaken in 1951-1952. The passage of over 30 years and the fact that the contracts were subject to the long-superseded Second Edition of the ICE Conditions of Contract means that the decision cannot be applied to particular circumstances arising out of the Fifth Edition of the ICE Conditions of Contract. Nevertheless, it remains useful for some general points on the presentation of claims.

The facts were these: the Engineer issued certain drawings which detailed the arrangement of drains in common trench and effectively specified a particular method of working; the contractor contended these were a late instruction and amounted to a variation of the works but the Engineer stated that they were only an amplification for the contractor's benefit of matters already fully covered in the contract documents; about five months later (when less than 10 per cent of the work in question had been executed) the contractor gave notice of a claim and began to submit accounts for reimbursement. As the case proceeded through the Courts two important points of continuing relevance were made in the judgements.

On what is required of a valid notice of claim:

'(It) does not mean a precisely formulated claim with full details, but it must be such a notice as will enable the party to whom it is given to take steps to

meet the claim by preparing and obtaining appropriate evidence for that purpose.'

On the timing of a claim submission:

'The Contractor would not be debarred for ever from claiming a particular item merely because through mistake or mis-apprehension or forgetfulness he failed to include the claim in the appropriate account... (nor because of) a merely technical and not substantial failure to adopt the proper procedure.'

The main requirement, therefore, is good faith on the part of the contractor who can be excused a limited amount of inefficiency and ignorance but not any 'tactical' manoeuvring designed to put the employer at a disadvantage.

Clause 12 claims

Although clause 52(4) provides a framework for all claims, there is one exceptional case which, whilst still subject to this general procedure, has its own supplementary provisions: a clause 12 claim.

This clause allows the contractor to claim when, *in his opinion*, extra costs are incurred because of adverse physical conditions or artificial obstructions 'which could not reasonably have been foreseen by an experienced Contractor'. Ground conditions are the most frequent source of clause 12 claims, the usual assertion being that the soils information in the contract documents, upon which the contractor is said to have relied when preparing his estimate and programme, was either inaccurate or insufficient.

It is essential to bear in mind the special nature of the clause 12 claim. It is unique because it can be made on the contractor's initiative without any positive act (such as the issue of a variation order) or omission (such as failure to give possession of part of the site) by the Engineer or his representative. For this reason special requirements as to notice, counter-notice and procedure are incorporated in the clause as a safeguard to ensure that the employer's interests are not prejudiced by a concealed or delayed claim.

The requirements for notice in clause 12 are strict and only in exceptional circumstances can they be waived (Tersons, it should be noted, was not a clause 12 claim). In particular, the contractual rights of the employer depend upon the notice being specific and contemporary. The wording of the clause describes a notice given at the time the condition or obstruction is detected. For example, the contractor is required to give details of the measures 'he is taking or is proposing to take', furthermore, the resident engineer is assumed to have the opportunity to obtain an estimate or consider a suspension. Clause 12, therefore, implies prompt notification, and its provisions, together with the general requirements set out in clause 52(4), must be implemented at once. In the case of any delay, it will be for the contractor to show that it was reasonable, in all the circumstances, for him to remain unaware of the adverse condition or obstruction for that length of time.

Two further provisions elsewhere in the ICE Conditions of Contract, but specifically aimed at clause 12, confirm the particular importance of this clause. Firstly, the authority to accept the existence of unforeseen physical conditions or artificial obstructions and to take them into account in assessing extra costs or delays is reserved strictly to the Engineer for the works and cannot be delegated. Secondly, the procedure for settling contractual disputes by arbitration contains a special provision entitling both the contractor and the employer to seek an arbitrator's award before completion of the works, but only in respect of a dispute arising under clause 12.

The process set out in the ICE Conditions of Contract requires the following steps to be taken:

(1) The contractor serves a notice of claim, specifying the conditions or obstruction encountered.
(2) With the notice, or as soon as possible afterwards, the contractor must provide details of the problem, his proposals to overcome it and the extent of the delays or difficulties he expects to suffer.
(3) The resident engineer may respond by approving the contractor's proposals with or without modifications and with or without an estimate of their cost; he also has the option of giving his own instructions on how to deal with the problem.
(4) If the resident engineer decides that the difficulty ought reasonably to have been foreseen, then he must notify

the contractor at once.

(5) Where the principle of the claim is accepted, the Engineer (this power is not delegable) must take into account any delays in determining the contractor's entitlement to an extension of time, and assess and certify any additional costs.

The following examples illustrate some typical clause 12 situations.

The contract for a major sewage outfall involved a section of tunnelling. The contract documents included details of a site investigation which indicated sound rock and showed no water strikes. When the heading had been driven over about half the length, badly fissured rock was encountered and very severe ground water flows. The contractor had to discontinue conventional tunnelling and adopt compressed air techniques.

The contractor on a highway project commenced excavation in a deep cutting where the earthworks drawings showed the majority of the material to be suitable fill. As work proceeded it became apparent that almost half the cutting would have to be classified as unsuitable and run to tip. The agent had to purchase a large quantity of imported fill, negotiate a licence to tip the unwanted material on adjacent land and reorganise the whole of his earthworks programme.

An overhead high-voltage electricity line crossed part of a site where extensive piling operations were required. The contract documents showed that the cables would be removed before piling commenced but this proved impossible to achieve. Although some operations beneath the power lines could continue, the movements of the piling rigs about the site were restricted and certain piles had to be driven out of sequence and others delayed.

A very contentious clause 12 claim can arise when it becomes apparent during the course of the works that the design has not fully taken into account a certain physical condition and has to be modified. The contractor is likely to argue strongly that, as the designer had failed to foresee the condition, he should not be expected to have taken it into account in his tender. This argument may not be valid. When the condition is adequately described elsewhere in the contract documents, the contractor has no grounds for using the designer's error as an excuse for his own failure to provide a 'sufficient' tender as required by clause 11 of the ICE Conditions of Contract. If, for example, a subway is to be constructed below existing ground level and the site investigation boreholes clearly indicate that water will be encountered, the omission of adequate drainage arrangements from the design of the permanent works does not entitle the contractor to claim under clause 12 when the excavation floods. He will, of course, be entitled under clause 52 for payment for the variation which covers the installation of the missing drains.

Assessment

In assessing the contractor's submission, the site staff follow a three-stage process. First, the principle upon which the claim is based is examined to determine whether it has any contractual validity. In its simplest form, this means checking whether the ICE Conditions of Contract place the liability for the event or action giving rise to the claim on the employer. Thus a contractor seeking reimbursement for losses arising out of an error in his setting-out cannot succeed in establishing any principle on which payment can be made, whereas the late issue of drawings or instructions does provide a basis for a legitimate claim. The issues are not always so clearly defined, however, and the existence of a principle may depend upon matters of engineering judgement. In a clause 12 claim the fundamental question is rarely the simple one of whether or not an artificial obstruction has been encountered but rather, could the obstruction reasonably have been foreseen by an experienced contractor?

If a principle has been identified, the assessment can proceed to the second stage, which is to investigate how and to what extent the grounds of the claim have caused the contractor actual loss or delay. The influence of other factors, for which the employer is not liable, must be assessed to ensure that they are eliminated from the evaluation. When, for example, the contractor does not obtain possession of a part of the site on time it may nevertheless be some other event, such as the default of a supplier or the failure to provide suitable haul roads, which effectively delays the operation and results in extra costs. In this instance, the principle for a claim exists but generates no payment, for the chain of cause and effect has been broken by the intervention of a more serious interruption to progress.

Having established that the employer is liable for the cause of the contractor's direct or indirect losses, the third and final stage is to evaluate the payment. A full accounting of all the contractor's relevant costs is essential to this part of the process and every item must be checked against both the site records to confirm its accuracy and the 'cause and effect' reasoning in the submission to affirm its connection with the principle of the claim. When the contractor is seeking additional payment to supplement an existing bill rate, it will be necessary to have verifiable evidence of how the original rate was built up at tender stage.

When the resident engineer considers that the claim is well-founded in principle and properly established as the cause of extra costs, the evaluation stage is relatively straightforward. Matters of fact should all be capable of resolution, provided comprehensive site records are available. The two major difficulties most frequently encountered are over the release of the tender 'build-ups' (which may limit the contractor's room for manoeuvre on other claims), and the extent to which the extra costs or delays ought to have been minimised by the contractor's efforts (the obligation to mitigate the effects of any loss or disruption is a contractual fact, its discharge one of fine engineering judgement). In any event, the resident engineer is able to make some payment provided the contractor has presented an interim account and acceptable supporting particulars.

The process becomes much more difficult when, at the first or second stage of the assessment, the resident engineer finds himself unable to agree to the contractor's claim. Although it is possible for a claim to fail on one specific point, it is more likely that, in the examination of the many statements which the contractor has made in building up his case, the resident engineer disagrees with his interpretations, detects errors and omissions, or reaches different conclusions. It is a gradual process, which requires a line-by-line analysis of the submission, and the evidence for and against the contractor's case has to be carefully weighed. To ensure that no subjective assessments are made on an unsupported 'feeling' for the weakness or otherwise of the arguments in the submission, the details of the assessment must be recorded. This disciplines the site staff and also provides an invaluable reference for subsequent re-examination if the matter continues as a dispute.

A most effective technique is to prepare a commentary by binding a copy of the claim 'in parallel' with the assessment so that, when opened, the left-hand page is from the contractor's submission and the right-hand page carries the corresponding notes prepared by the site staff. To make reference easier particular points in the submission can be numbered. Figure 16 is an extract from the commentary on an earthworks claim, and illustrates both the format and the content.

In conducting the assessment, the resident engineer must give due consideration to the question of 'sufficiency of tender'. He must ask himself whether, at tender stage, it would have been reasonable to expect an experienced contractor to have foreseen the circumstances and risks upon which the claim is based and allowed for them in his estimate. It is important that the answer should take into account the accuracy and completeness of the information presented, with a considerable amount of implied authority, in the tender documents and the relatively short time available for tenderers to conduct their own investigations and site inspection. It is equally necessary to recognise that the employer sets out to purchase expertise and not mistakes. The fact that *this* particular contractor failed to cover some risk or circumstance in his tender is not relevant, for liability depends upon what an 'experienced Contractor' would have done in that situation. Contracting is a risk business, in which the contractors take their profits or stand their losses as the terms of the contract may dictate.

In conducting the assessment of a claim submission, the site staff must strike a careful balance between obstructing the contractor and assisting him. If his investigation shows that there is a fundamental error of fact in the narrative, that certain dimensions are incorrect or that some essential confirmatory evidence has been omitted but that the principle of the claim remains sound, the resident engineer must point out the deficiencies and allow the agent the opportunity to re-submit. Where clarification or amplification is required, specific questions can be put to the agent. On the other hand, if the submission does not establish a principle for payment or fails to demonstrate how the alleged extra costs or delays have arisen, the resident engineer need only state in general terms the reasons for its rejection. It is not part of the supervisory role to formulate claims on the contractor's behalf, nor to provide advice on what will produce the best settlement — these facilities ought to be available within the contractor's own organisation, or, if not, can be obtained (at a price) from the many specialist consultants operating in this field.

Disputes and the Engineer's decision

In discussing the role of the Engineer (in Chapter 2) his duty to 'act fairly as between the parties' was stressed. So also was his function as an adjudicator. These responsibilities assume great importance when the contractor and the employer are in dispute.

A formal dispute is usually the result of the resident engineer's refusal to certify, in whole or in part, some item of measured work or claim for additional costs and the Engineer's subsequent confirmation of this action. In contractual terms, the contractor's dispute is not with the Engineer but with the employer, and clause 66 of the ICE Conditions of Contract sets out the procedure by which such disputes are referred to, and settled by, the Engineer. The result is commonly known as an 'Engineer's decision'. The procedural formalities must be scrupulously followed, for there are time limits to be observed on which the parties' rights of subsequent action may depend. Thus the contractor (or, more rarely, the employer, for both may refer a dispute to the Engineer) must state clearly that he is seeking a decision under clause 66. Similarly, the Engineer's response, which must be delivered within three months, must be described as 'a decision in accordance with clause 66' or in similar unequivocal terms. The contractor must proceed with the works and, whether he agrees with it or not, give effect to the Engineer's decision.

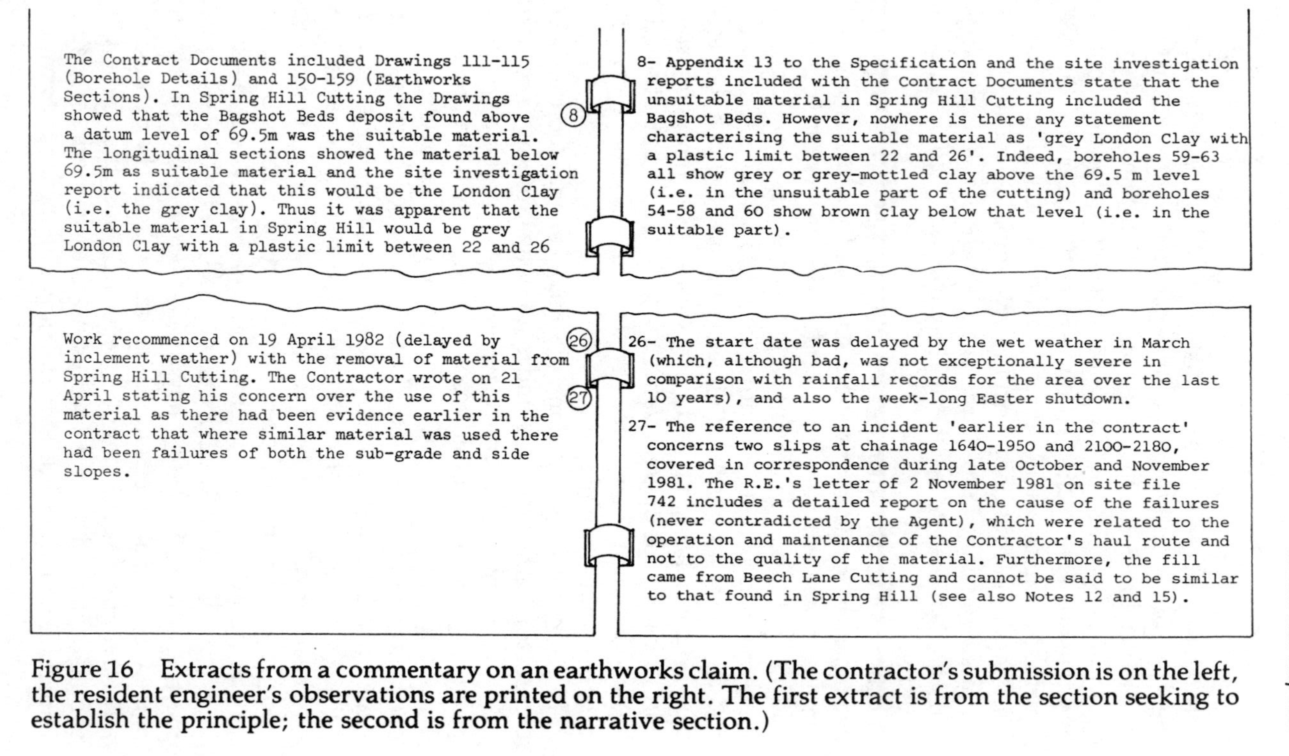

The Contract Documents included Drawings 111-115 (Borehole Details) and 150-159 (Earthworks Sections). In Spring Hill Cutting the Drawings showed that the Bagshot Beds deposit found above a datum level of 69.5m was the suitable material. The longitudinal sections showed the material below 69.5m as suitable material and the site investigation report indicated that this would be the London Clay (i.e. the grey clay). Thus it was apparent that the suitable material in Spring Hill would be grey London Clay with a plastic limit between 22 and 26 (8)

8- Appendix 13 to the Specification and the site investigation reports included with the Contract Documents state that the unsuitable material in Spring Hill Cutting included the Bagshot Beds. However, nowhere is there any statement characterising the suitable material as 'grey London Clay with a plastic limit between 22 and 26'. Indeed, boreholes 59-63 all show grey or grey-mottled clay above the 69.5 m level (i.e. in the unsuitable part of the cutting) and boreholes 54-58 and 60 show brown clay below that level (i.e. in the suitable part).

Work recommenced on 19 April 1982 (delayed by inclement weather) (26) with the removal of material from Spring Hill Cutting. The Contractor wrote on 21 April stating his concern over the use of this material as there had been evidence earlier in the contract (27) that where similar material was used there had been failures of both the sub-grade and side slopes.

26- The start date was delayed by the wet weather in March (which, although bad, was not exceptionally severe in comparison with rainfall records for the area over the last 10 years), and also the week-long Easter shutdown.

27- The reference to an incident 'earlier in the contract' concerns two slips at chainage 1640-1950 and 2100-2180, covered in correspondence during late October and November 1981. The R.E.'s letter of 2 November 1981 on site file 742 includes a detailed report on the cause of the failures (never contradicted by the Agent), which were related to the operation and maintenance of the Contractor's haul route and not to the quality of the material. Furthermore, the fill came from Beech Lane Cutting and cannot be said to be similar to that found in Spring Hill (see also Notes 12 and 15).

Figure 16 Extracts from a commentary on an earthworks claim. (The contractor's submission is on the left, the resident engineer's observations are printed on the right. The first extract is from the section seeking to establish the principle; the second is from the narrative section.)

Two vitally important factors govern the process by which the Engineer reaches his decision: firstly, there can be no delegation and the Engineer must exercise his personal judgement; secondly, although it is understood that he is not acting in a judicial role, the Engineer must conduct himself 'fairly as between the parties'. A clause 66 decision is a serious matter and must be approached with the respect it deserves, but the general principles are no different from those which the resident engineer would apply to his own initial assessment of a claim submission, indeed, the site staff should always carry out that assessment as if they were making an Engineer's decision.

Having had the dispute referred to him, the Engineer examines the contractor's submission and any statement the employer may wish him to consider. He assesses the contents and also collects and examines any other evidence which is, in his opinion, relevant, this normally includes the site records and any observations or detailed commentary which the resident engineer has prepared. The views of his representative will, of course, be of great value to the Engineer, but although he may endorse them after proper deliberation, he cannot simply 'rubber stamp' them.

In coming to his decision, the Engineer may consult the contractor, although he is under no obligation to do so. He may do so by correspondence, seeking an explanation of a point or requesting certain records, or he may call a meeting. This latter course is often urged on the Engineer by the contractor and must be embarked upon only with caution; for whilst it may prove a useful means of clarifying issues and cutting through obscure contractual dissertations to reveal the real heart of the dispute it may just as well lead to confusion and prejudice a fair settlement. Statements can easily be taken out of context, misunderstood or misconstrued. 'Leading questions' can produce unprepared and ill-advised answers. If meetings are to be held, they should be kept strictly to an agreed agenda, all questions of substance should be notified in advance and formal minutes recorded, and the employer should be invited to send a representative (the Engineer, being impartial, could not fulfil this role). For all the benefits which meetings are said to bring, and the value of these 'man to man discussions' is often highly praised by those seeking to arrange them, it must be said that the English language is a very flexible instrument and it is difficult to conceive of what can be said around a table which cannot be reduced to writing. One aspect of the Engineer's non-judicial status is that he can be held liable in negligence for his decision. It is, therefore, unwise to take into account evidence or argument which is not on the record nor is it 'fair' to allow the contractor the opportunity of a special hearing without affording the employer similar facilities.

The ICE Conditions of Contract do not state what form the decision must take, except that it must be in writing, and there is no specific requirement that it should give reasons. Few, if any, Engineers (or commentators on contract law) would suggest that the decision should be a detailed 'judgement', setting out in full the process of examining and interpreting the evidence which resulted in the decision. This would serve no purpose beyond producing the conditions for a 'chain reaction' of subsidiary disputes over points of detail, which would contribute nothing towards the major issue of whether or not the decision is satisfactory to both parties. Some Engineers might argue for the opposite extreme, a straight 'yes or no' answer. This approach would make it impossible to sue the Engineer for a negligent or capricious decision, but that would be its only benefit for such a response would be of no help to the parties in deciding between acceptance and further proceedings.

Arbitrators are now generally expected, but not bound, to give reasoned decisions, and Engineers should adopt a similar standard. For guidance, they should consider the remarks of Lord Donaldson who advised arbitrators that their reasons should 'set out what, on their view of the evidence, did or did not happen and should explain succinctly why, in the light of what happened, they have reached their decision and what that decision is.' Hence, an Engineer in his decision should make it clear on what principles he has based his findings. To take the rejection of a clause 12 claim as an example, the decision should state succinctly on which of the following points the claim has failed:

(1) notice and procedure;
(2) principle (that is, the presence of an unforeseen condition or obstruction);
(3) the link between principle and actual loss or delay;
(4) the substantiation of the alleged extra costs or delays.

Arbitration

Clause 66 states that the Engineer's decision is 'final and binding', and so it is as far as the Engineer is concerned. There is, however, one further stage provided for in the contract. If the contractor or employer is dissatisfied with the Engineer's decision, he may, up to three months after notification of the decision, elect to have the matter referred to arbitration.

Although the dissatisfied party has shown his intention to pursue the dispute, he must comply with the Engineer's decision and continue with the works. No arbitration proceedings can take place until after completion and any attempt in the meantime to defy or frustrate the decision would be a serious breach of contract. An exception is made in the case of clause 12 disputes, for which there is special provision in clause 66(2) for interim arbitration as construction proceeds.

The arbitrator may or may not be a lawyer, he could be a practising civil engineer, but he will be skilled in dealing with construction disputes and have personal knowledge of the activities and customs of the industry. He is empowered to make orders compelling each party to define the points of their case and disclose all their documentary evidence before calling a hearing at which he will consider the arguments put forward by the employer and contractor and make an award. The award is final and can only be questioned in the courts on a point of law. The proceedings and the award are private and so arbitrators are not bound by precedent. This allows complete flexibility, albeit with some loss of consistency. Arbitrators have acquired something of a reputation for leniency (contractors would substitute 'realism') in their views on certain aspects of contractual procedure and liability particularly in respect of clause 12 claims.

A number of text books and commentaries deal in detail with arbitration, some — though by no means all — giving useful insights into the faults as well as the benefits of the process in its current form. The Institution of Civil Engineers' 'Guide to ICE Arbitration Procedure' is also helpful as it not only provides guidelines for the conduct of proceedings but also gives a good idea of how the system works.

The procedure is designed to be quicker and cheaper than an action in the courts, because of the less formal, more flexible rules and the arbitrator's technical expertise. However, many arbitrations are very complex affairs, complete with barristers, solicitors, expert witnesses and reams of evidence in circulation and, apart from the absence of ceremonial, are almost indistinguishable from court proceedings.

They certainly can be extremely expensive and for this reason many disputes are settled during the preliminary proceedings and before the hearing proper gets under way. Such a settlement would be a matter between the employer and contractor, for the Engineer has no part to play in the conduct of an arbitration other than as the source of much of the evidence, site records, contemporary reports etc., and possibly as a witness. Once he has given a decision, the Engineer has no further powers to exercise in the resolution of a dispute.

Because the stakes are so high in modern contracting, arbitration is no longer as rare as it used to be. Site staff who are called to attend a hearing may reflect upon the inordinate amount of time and effort expended on arriving at an independent judgement on an issue which has already been considered by an impartial professional expert. They may also recognise the great value of a sound decision delivered with confidence by an Engineer whose staff have conducted the supervision of the contract with sufficient skill to have earned the respect of employer and contractor.

References to the ICE Conditions of Contract (5th edition)

For summary of main clauses see pages 13–16.

Bibliography

Arbitration procedure (1983). Institution of Civil Engineers, London, 1983.

Civil engineering standard method of measurement. Institution of Civil Engineers, London, 1976.

Conditions of contract and forms of tender, agreement and bond for use in connection with works of civil engineering construction (5th edn). Institution of Civil Engineers, London, 1973.

Constitution and working rule agreement. Civil Engineering Construction Conciliation Board of Great Britain, London, 1984.

Form of sub-contract for use in conjunction with the ICE Conditions of Contract. Federation of Civil Engineering Contractors, London, 1973.

Method of measurement for road and bridge works. Department of Transport, Scottish Development Department, Welsh Office. HMSO, London, 1977.

Model form of agreement 'A' between a client and consulting engineers for design and supervision of works of civil engineering construction. Association of Consulting Engineers, London, 1963, 1979.

Notes for guidance and library of standard item descriptions for the preparation of bills of quantities for road and bridge works. Department of Transport, Scottish Development Department, Welsh Office, HMSO, London, 1978.

Notes for guidance on the specification for road and bridge works. Department of Transport, Scottish Development Department, Welsh Office, HMSO, London, 1977.

Price adjustment formulae for construction contracts: monthly bulletin of indices. Department of the Environment, HMSO, London.

Schedule of dayworks carried out incidental to contract work. Federation of Civil Engineering Contractors, London, 1983.

Specification for road and bridge works. Department of Transport, Scottish Development Department, Welsh Office, HMSO, London, 1976.

Index